Jonathan Schulte

Multi-field modeling and simulation of fiber-reinforced polymers

Schriftenreihe des Instituts für Mechanik
Karlsruher Institut für Technologie (KIT)

Band 9

Herausgeber:
Prof. Dr.-Ing. habil. Peter Betsch
Prof. Dr.-Ing. habil. Thomas Seelig

Eine Übersicht aller bisher in dieser Schriftenreihe erschienenen Bände
finden Sie am Ende des Buches.

Multi-field modeling and simulation of fiber-reinforced polymers

by
Jonathan Schulte

KIT Scientific Publishing

Dissertation, Universität Siegen
Naturwissenschaftlich-Technische Fakultät
Tag der mündlichen Prüfung: 25. Oktober 2021
Betreuer und erster Gutachter: Prof. Dr.-Ing. habil. Christian Hesch, Universität Siegen
Zweiter Gutachter: Priv.-Doz. Dr.-Ing. habil. Fadi Aldakheel, Leibniz Universität Hannover

Impressum

KIT Scientific Publishing

Karlsruher Institut für Technologie (KIT)
KIT Scientific Publishing
Straße am Forum 2
D-76131 Karlsruhe

KIT Scientific Publishing is a registered trademark
of Karlsruhe Institute of Technology.
Reprint using the book cover is not allowed.

www.ksp.kit.edu

Print on Demand 2023 – Gedruckt auf FSC-zertifiziertem Papier

ISSN 2363-4936
ISBN 978-3-7315-1251-6
DOI 10.5445/KSP/1000151847

Abstract

In recent years, fiber-reinforced materials have risen to prominence in a variety of applications spanning from aircraft and automobile engineering to medical technology and biological materials. Despite the scientific progress in understanding this class of materials achieved in diverse research contributions, the precise modeling and simulation of macroscopic systems with dedicated microstructure up to ultimate fracture remains an open research field in mechanics.
The present work presents a novel numerical approach to analyze the thermomechanical behavior within composite materials including the inelastic regime up to final failure. Therefore, a second-gradient theory is combined with phase-field methods to fracture. In particular, we assume that the polymeric matrix material undergoes ductile fracture, whereas continuously embedded fibers undergo brittle fracture as it is typical e.g. for roving glass reinforced thermoplastics. A hybrid phase-field approach is developed and applied in conjunction with a modified Gurson-Tvergaard-Needleman GTN-type plasticity model accounting for a temperature-dependent growth of voids on microscale. The mechanical response of the arising microstructure of the woven fabric gives rise to additional higher-order terms, representing homogenized bending contributions of the fibers. Eventually, a series of tests is conducted for this physically comprehensive multifield formulation to investigate various types and sequences of failure within long fiber-reinforced polymers.

Kurzfassung

In den letzten Jahren haben faserverstärkte Werkstoffe in einer Vielzahl von Anwendungen vom Flugzeug- und Automobilbau über die Medizintechnik bis hin zu biologischen Materialien eine immense Bedeutung gewonnen. Trotz der wissenschaftlichen Fortschritte im Verständnis dieser Materialklassen, die in diversen Forschungsbeiträgen erzielt wurden, bleibt die präzise Modellierung und Simulation makroskopischer Systeme mit dedizierter Mikrostruktur bis hin zum endgültigen Bruch ein offenes Forschungsfeld in der Mechanik. Die vorliegende Arbeit stellt einen neuartigen numerischen Ansatz vor, um das thermomechanische Verhalten in Verbundwerkstoffen einschließlich des inelastischen Regimes bis zum endgültigen Versagen zu analysieren. Dazu wird eine zweite Gradienten Theorie mit Phasenfeldmethoden für den Bruch kombiniert. Insbesondere wird davon ausgegangen, dass das polymere Matrixmaterial einen duktilen Bruch erfährt, während kontinuierlich eingebettete Fasern einen Sprödbruch erfahren, wie er z.B. für glasfaserverstärkte Thermoplaste typisch ist. Es wird ein hybrider Phasenfeldansatz entwickelt und in Verbindung mit einem modifizierten Gurson-Tvergaard-Needleman GTN-Plastizitätsmodell angewendet, das ein temperaturabhängiges Wachstum von Hohlräumen auf der Mikroskala berücksichtigt. Die mechanische Charakterisierung der entstehenden Mikrostruktur des Gewebes führt zu zusätzlichen Termen höherer Ordnung, die homogenisierte Biegebeiträge der Fasern darstellen. Schließlich wird für diese physikalisch umfassende Mehrfeldformulierung eine Versuchsreihe durchgeführt, um verschiedene Versagensarten und -folgen in langfaserverstärkten Polymeren zu untersuchen.

Acknowledgements

The research presented in this contribution was carried out during my time as student and research assistant at the Chair of Computational Mechanics at the University of Siegen from 2016 to 2021. During the past years, I had the great privilege to meet many people who helped me to mature both personally and scientifically and thus enabled me to successfully complete this work. I would like to take this opportunity to express my sincere gratitude to all those who have supported me in various ways.

First and foremost, I would like to thank my advisor Professor Christian Hesch for his endless support, for his motivation and inspiring scientific passion and most of all for the extremely friendly and productive working atmosphere. I have thoroughly enjoyed working at the Chair of Computational Mechanics over the past years and will always remember this great, formative time.

I would also like to thank Doctor Fadi Aldakheel for his sincere interest in my work and for his willingness to serve as a co-referee. A big thank you also goes to Professor Peter Kraemer and the chairman of the examination board Professor Xiangfan Fang.

Next, I would like to thank all my colleagues for the pleasant working atmosphere, for many fruitful discussions, and for the great time spent together: Harald Schmitz, Stefan Schuss, Felix Schmidt, Melanie Krüger, Annika Ippach, Alex Härtel, and last but not least Maik Dittmann, who introduced me to the world of numerical mechanics during my time as a student assistant and deeply inspired me with his incredible work ethic.

Last and most importantly, I thank my beloved wife Richmelle, our daughters Soraya, Zoé, and Séphora, and our son Zayan from the bottom of my heart. Without their inspiration, support, patience and love, this work would not have been possible.

Siegen, December 2021 Jonathan Schulte

Publications

This thesis is based on the following peer-reviewed journal articles which were published during the preparation of the thesis:

M. Dittmann, J. Schulte, F. Schmidt and C. Hesch: A strain-gradient formulation for fiber-reinforced polymers: Hybrid phase-field model for porous-ductile fracture, *Computational Mechanics*, 67:1747–1768, 2021.

J. Schulte, M. Dittmann, S.R. Eugster, S. Hesch, F. dell'Isola and C. Hesch: Isogeometric analysis of fiber-reinforced composites using Kirchhoff-Love shell elements, *Computer Methods in Applied Mechanics and Engineering*, 362:112845, 2020.

M. Dittmann, F. Aldakheel, J. Schulte, F. Schmidt, M. Krüger, P. Wriggers and C. Hesch: Phase-field modeling of porous-ductile fracture in non-linear thermo-elasto-plastic solids, *Computer Methods in Applied Mechanics and Engineering*, 361:112730, 2020.

M. Dittmann, F. Aldakheel, J. Schulte, P. Wriggers and C. Hesch: Variational phase-field formulation of non-linear ductile fracture, *Computer Methods in Applied Mechanics and Engineering*, 342:71-94, 2018.

For the journal articles mentioned above, which were prepared in collaboration with one or more co-authors, the author of this thesis contributed fundamental aspects with regard to the outline of the theory, carried out the numerical implementations, evaluated the simulations and wrote the articles.

Contents

List of Figures

List of Tables

1 Introduction

Composite materials have played a significant part in human development throughout history. Thus, the combination of clay and straw for the production of bricks is known from the excavation of the city of Çatalhöyük [77], the first documented city, dated back to about 8000 BC. Whilst straw improves the tensile capacity of clay, it will also act to assist in keeping the moisture content of the material more constant throughout, leading to less drying shrinkage cracks.

In the development of polymeric fiber composites, the low density and high strength fiber material, typically glass or carbon, is combined with a polymer matrix in order to create geometrically defined components. The matrix serves not only to shape the material, but also to transfer loads to the fibers and to protect the fibers from environmental influences and abrasive wear caused by mutual contact. An early example of the use of polymer matrix composites are the Micarta aircraft propellers from around 1920 onwards [71]. Due to their excellent cost-performance ratio as well as their adjustable mechanical properties, fiber-reinforced composites are nowadays used in a wide range of industrial applications. In recent years, new types of additive manufacturing machines have been developed and brought to market that are capable of designing customized composites by introducing fibers into the produced parts during the manufacturing process. Until now, however, the design of structural composite components has been governed by an incomplete knowledge, which is mainly based on empiricism.

However, the dramatic increase in computing power over the past decades has made possible more and more realistic simulations of complex materials. We can now resolve sophisticated micro structures in a multi-physical environment using brute-force computations within a continuum mechanical framework. New challenges arise for the implementation of composite materials in a suitable simulation environment in order to obtain realistic predictions of the material behavior. In particular, the distribution and orientation of the fibers on the microscale have a significant influence on the macroscale mechanics. Although of great interest to the industry as a whole, numerical prediction of thermomechanical macroscopic systems with dedicated microstructure behavior including arbitrary three-dimensional fracture patterns of fiber-reinforced polymers is currently subject to large uncertainties and still under research. This is a dramatic information deficit, since fiber-reinforced polymers are already in use in vital structures.

1.1 State of the art and objectives

1.1.1 Strain-gradient formulations for fiber-reinforced materials

Fiber-reinforced materials can be understood as having detailed microstructure because the length scales of the embedded fibers are small compared to the surrounding continuum. A very general framework for including the effects of microstructures within a continuum formulation was presented in the seminal and fundamental work on higher-order theories by Mindlin [89, 90], see also the work of Germain [53], Toupin [121, 122], and Eringen [21].

A number of publications establish a connection between the microstructure and the macroscopic formulation as generalized continua, see, among others [81, 49, 60, 105], see also [72, 67, 97, 98, 129] for more details on the mathematical structure. Specific mechanical problems such as elastic meshes have been treated by Steigmann et al. [118, 115, 117, 116], for application to panthographic structures see dell'Isola et al. [31, 32].

Within the present work a model of a woven fabric as presented in Steigmann [116] will be first be embedded into the Kirchhoff-Love shell theory and then be adapted to three-dimensional continua. Therein, the in-plane flexural resistance of the fibers is taken into account in addition to first-gradient anisotropic effects. We will demonstrate on the basis of experimental results that a classical Cauchy continuum theory without higher-gradient contributions can only be adapted to a specific fiber orientation and load case, but never independently on the orientation. In contrast, the proposed formulation as generalized continuum with higher-gradient contributions allows for an independent modeling without recalibration of the material for the specific fiber direction. Further discussions on higher-order contributions for the constitutive modeling of composites have been presented in Asmanoglo and Menzel [10, 11], using the framework as provided in the preliminary work of Spencer and Soldatos [114] and Soldatos [113]. Note that we will show here, that the formulation as proposed by dell'Isola et al. [32] can be recast into the formulation of Asmanoglo and Menzel [10, 11].

To obtain the corresponding constitutive information, classical homogenization methods are often applied, which are usually limited to the first gradient anisotropic contributions. In particular, full-field homogenizations are usually based on a unit cell discretized with finite elements (see, e.g., in Fish [46] for a comprehensive review of multiscale methods) and have been applied to composites, including in Wang et al. [125], see also the references in Mishnaevsky [93]. Basic ideas on the homogenization of continua with micromorphic mesostructure can be found in Hirschberger [59]. Recent studies on the homogenization of continua with micromorphic properties can be found in Hutter [65].

For the spatial discretization, B-spline based shape functions are used, which originate from the field of Computer Aided Design (CAD). The corresponding finite element framework is commonly known as isogeometric analysis (IGA), see Hughes et al. [64] and Cottrell et al. [26]. IGA allows to control the differentiability of the shape functions and is therefore suitable for higher-gradient continua, see Fischer et al. [45] for applications to strain-gradient elasticity. Kirchhoff-Love shell formulations have also been analyzed in recent years using the concept of IGA, see Kiendl et al. [69, 70], among others.

1.1.2 Phase-field approach to fracture

Along this line, Choo and Sun [22] have developed a framework that combines a phase-field approach to fracture with a pressure-sensitive plasticity for modeling brittle fracture to ductile flow in geological materials. A coupled crystal-plasticity-phase-field fracture formulation based on experimental observations was presented in Diehl et al. [33]. A gradient-extended plasticity model for overcoming mesh sensitivities, such as that presented in De Borst and Mühlhaus [30], Fleck et al. [47], Menzel and Steinmann [82], Geers [52], Anand et al. [9], Dimitrijevic and Hackl [34], Miehe et al. [87], Aldakheel [3, 5], Wulfinghoff and Böhlke [127], and Forest [48] was applied to phase-field breaking problems in Aldakheel et al. [4], Miehe et al. [84], Aldakheel [2], and Dittmann et al. [35].

A variety of phenomenological and micromechanical approaches exist in the literature for modeling the diverse damage mechanisms in the polymers used here as matrix materials. To describe such phenomena, material behavior at the microscale must be included in the continuum formulation. In particular, the growth of microvoids prior to ultimate fracture must be considered at the macroscale, as rooted in the seminal work of Gurson [56, 57]. In this work, a macroscopic flow surface was developed by homogenizing a porous representative volume element with assumed rigid plastic flow, which degrades with increasing void fraction. Later, this model was used by Tvergaard [123], Tvergaard and Needleman [124], Needleman and Tvergaard [96], Leblond et al. [74], Nahshon and Hutchinson [95], Xue et al. [128], Li et al. [76], and Huespe et al. [62] to account for damage growth, extending the flow criterion function by introducing new material parameters to account for nucleation and coalescence effects. In Hutter et al. [63] and Reusch et al. [103, 104], nonlocal Gurson models were presented to overcome the nonphysical mesh sensitivity in the softening materials. Recent works on the application of this model to polyamide, as commonly used for fiber-reinforced thermoplastics, can be found, for example, in Selles et al. [109] and Cayzac et al. [20]. Thermomechanical extensions of the Gurson-type model with brittle crack propagation in thermoelastic solids were shown in Dittmann et al. [39] and Miehe et al. [86]. An extension towards finite-strain thermo-poros-plasticity based on the phase-field approach was recently developed in Dittmann et al. [36].

In this contribution, we introduce a novel hybrid phase-field model to simulate failure mechanisms in fiber-reinforced materials. In particular, we assume that the polymeric matrix material undergoes ductile fracture, whereas continuously embedded fibers undergo brittle fracture. The matrix and fiber material are associated with indepedent phase-fields. A modified Gurson-Tvergaard-Needleman GTN-type plasticity model accounting for a temperature-dependent growth of voids on microscale is furthermore applied to the polymeric matrix material. Eventually, a series of tests is conducted for this physically comprehensive multifield formulation to investigate different kinds and sequences of failure within long fiber-reinforced polymers.

1.2 Outline of the thesis

The outline of the thesis is as follows. Chapter 2 introduces the basic notations and concepts of classical continuum mechanics used in this thesis. The approach to simulate porous-ductile fracture in isotropic thermo-elasto-plastic solids subjected to large deformations is then presented in Chapter 3. Chapter 4 provides an introduction to generalized continuum mechanics for microstructured materials. A very specific generalized continuum formulation for fiber materials is then outlined in Chapter 5. The complete framework for modeling and simulation of fiber-reinforced polymers, including the inelastic regime up to ultimate failure, is presented in Chapter 6. Finally, conclusions are drawn in Chapter 7.

2 Fundamentals

This chapter introduces the basic notations and concepts of this work. After recalling some tensor operations, the principle equations of non-linear continuum mechanics are presented. Next, the kinematics of Kirchhoff-Love shells are outlined. Finally, the phase-field approach to non-linear brittle fracture and the isogeometric concept for finite element analysis are introduced.

2.1 Tensor notations

In order to provide a clear representation of mathematical operations used therein we recall and define some specific tensor operations. Let $\boldsymbol{a} \in \mathbb{R}^3$ be a vector, $\boldsymbol{A} \in \mathbb{R}^3 \times \mathbb{R}^3$ a second-order tensor and $\boldsymbol{\mathfrak{A}} \in \mathbb{R}^3 \times \mathbb{R}^3 \times \mathbb{R}^3$ a third-order tensor.

We denote the symmetric, skew-symmetric, spheric, and deviatoric part of a second-order tensor, respectively, as

$$\begin{aligned} \operatorname{sym}(\boldsymbol{A}) &= \frac{1}{2}(\boldsymbol{A} + \boldsymbol{A}^T), \quad \operatorname{skew}(\boldsymbol{A}) = \frac{1}{2}(\boldsymbol{A} - \boldsymbol{A}^T), \\ \operatorname{sph}(\boldsymbol{A}) &= \frac{1}{3}\operatorname{tr}(\boldsymbol{A})\boldsymbol{I}, \qquad \operatorname{dev}(\boldsymbol{A}) = \boldsymbol{A} - \operatorname{sph}(\boldsymbol{A})\boldsymbol{I}, \end{aligned}$$

where $\boldsymbol{I}$ is the identity matrix.

In addition, we define the gradient with respect to the reference and current configuration $\nabla(\bullet)$ and $\nabla_x(\bullet)$ of a vector field as

$$[\nabla \boldsymbol{a}]_{iJ} = \frac{\partial[\boldsymbol{a}]_i}{\partial[\boldsymbol{X}]_J} \quad \text{and} \quad [\nabla_x \boldsymbol{a}]_{ij} = \frac{\partial[\boldsymbol{a}]_i}{\partial[\boldsymbol{x}]_j} \tag{2.1}$$

and of a second-order tensor field as

$$[\nabla \boldsymbol{A}]_{iJK} = \frac{\partial[\boldsymbol{A}]_{iJ}}{\partial[\boldsymbol{X}]_K} \quad \text{and} \quad [\nabla_x \boldsymbol{A}]_{iJk} = \frac{\partial[\boldsymbol{A}]_{iJ}}{\partial[\boldsymbol{x}]_k} \tag{2.2}$$

respectively. The divergence operator with respect to the reference configuration $\operatorname{Div}(\bullet)$ or $\nabla \cdot (\bullet)$ of a second-order tensor field $\boldsymbol{A}$ and third-order tensor field $\boldsymbol{\mathfrak{A}}$ is defined as

$$[\nabla \cdot \boldsymbol{A}]_i = \frac{\partial[\boldsymbol{A}]_{iJ}}{\partial[\boldsymbol{X}]_J} \quad \text{and} \quad [\nabla \cdot \boldsymbol{\mathfrak{A}}]_{iJ} = \frac{\partial[\boldsymbol{\mathfrak{A}}]_{iJK}}{\partial[\boldsymbol{X}]_K}, \tag{2.3}$$

respectively. Finally, the double contractions of a second-order and third-order tensor, i.e. $\boldsymbol{a} = \boldsymbol{A} : \mathfrak{A}$ and $\boldsymbol{b} = \mathfrak{A} : \boldsymbol{A}$, are defined as

$$[\boldsymbol{a}]_k = [\boldsymbol{A}]_{ij} : [\mathfrak{A}]_{ijk} \quad \text{and} \quad [\boldsymbol{b}]_i = [\mathfrak{A}]_{ijk} : [\boldsymbol{A}]_{jk}, \tag{2.4}$$

respectively.

2.2 Continuum mechanics

This section recalls the fundamental concepts of classical continuum mechanics at finite strains, also referred to as Cauchy-Boltzmann theory or first-gradient theory. First, the basic kinematical relations and deformation measures are introduced. Then, the concept of internal contact actions, i.e. stress and heat flux are presented. Finally, the physical balance equations are outlined, along with their associated local forms, i.e. the Eulerian equations of motion. A more extensive and systematical presentation of these arguments can be found in e.g. [23, 61, 110, 17].

2.2.1 Kinematics of finite deformation

Motion

A continuum body is a set of material points that are in bijective correspondence, at each instant of time, with the geometrical points of a region of the Euclidean space, denoted as $\mathbb{R}^3$. The abstraction of continuum body is an approximation since the actual structure of the materials is discontinuous, but it is necessary for performing the classical mathematical operation (e.g. differentiation) and widely validated through experiments.

Under the infinite possible configurations of the body, we call reference configuration $\mathcal{B}_0$ and current configuration $\mathcal{B}$ the configurations of the body at time $t = 0$ and $t \in \mathbb{R}^+$, respectively. The two configurations are equivalently called Lagrangian and Eulerian configuration and the material points are labeled as $\boldsymbol{X}$ and $\boldsymbol{x}$, respectively. The generic nonlinear transformation that maps the reference configuration into the current one defined as

$$\boldsymbol{x} = \boldsymbol{\varphi}(\boldsymbol{X}, t) \tag{2.5}$$

is called placement.

Deformation Gradient

The second-order tensor

$$\boldsymbol{F} = \nabla \boldsymbol{\varphi} \tag{2.6}$$

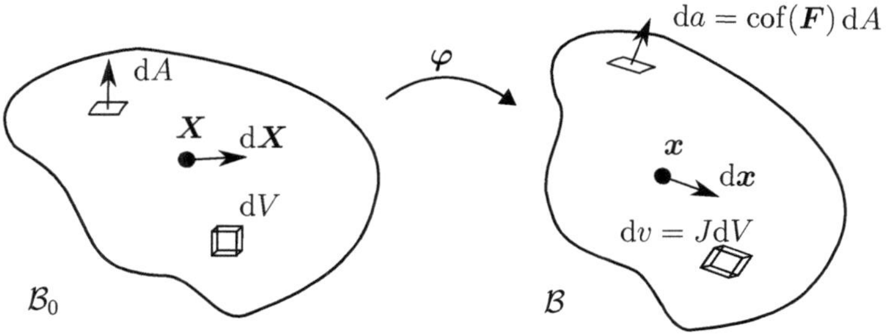

Figure 2.1: Motion of a material body in the reference configuration $\mathcal{B}_0$ with position $\boldsymbol{X}$ and current configuration $\mathcal{B}$ with position $\boldsymbol{x}$ at time t.

is called the deformation gradient. In the three-dimensional configuration, this tensor has nine independent components and characterizes the motion in the neighbor of a material point.

The determinant of the deformation gradient characterizes the measure of the volume change. To exclude self-penetration, the Jacobian determinant must always assume a value greater than zero

$$J = \det(\boldsymbol{F}) = \frac{1}{6}(\boldsymbol{F} \times \boldsymbol{F}) : \boldsymbol{F} > 0. \tag{2.7}$$

With the definition of the deformation gradient and its determinant, it is possible to define the deformation of a surface and volume element from the reference to the current configuration with

$$\mathrm{d}a = J\boldsymbol{F}^{-1}\,\mathrm{d}A = \operatorname{cof}(\boldsymbol{F})\,\mathrm{d}A = \frac{1}{2}(\boldsymbol{F} \times \boldsymbol{F})\,\mathrm{d}A \quad \text{and} \quad \mathrm{d}v = \det(\boldsymbol{F})\,\mathrm{d}V. \tag{2.8}$$

The Jacobian determinant J is thus a measure for the relative change in volume during a deformation. Note that the tensor cross product operation, see [29], was utilized for the first time in the domain of solid mechanics in Bonet et al. [16].

The deformation gradient can be decomposed in a multiplicative form through the polar decomposition, and one can obtain:

$$\boldsymbol{F} = \boldsymbol{R} \cdot \boldsymbol{U} = \boldsymbol{V} \cdot \boldsymbol{R} \tag{2.9}$$

which is called right and left polar decomposition. $\boldsymbol{R}$ is an orthogonal tensor (i.e. $\boldsymbol{R}^T = \boldsymbol{R}^{-1} \text{and} \det(\boldsymbol{R}) = 1$) and it represents a rotation, and $\boldsymbol{U}$ and $\boldsymbol{V}$ are symmetric and definite positive tensors. These two tensors are called right and left stretch tensor, respectively, and are linked via the rotation tensor through the formula

$$\boldsymbol{V} = \boldsymbol{R} \cdot \boldsymbol{U} \cdot \boldsymbol{R}^T. \tag{2.10}$$

Deformation Measures

Having defined the deformation gradient $\boldsymbol{F}$ and keeping in mind that it is not a suitable quantity for strain measures, we introduce the most common strain tensors related to either the material or the spatial description used in the field of nonlinear solid continuum mechanics. The two symmetric and positive definite strain tensors

$$\begin{aligned} \boldsymbol{C} &= \boldsymbol{U}^2 = \boldsymbol{F}^T\boldsymbol{F} \\ \boldsymbol{b} &= \boldsymbol{R}^2 = \boldsymbol{F}\boldsymbol{F}^T \end{aligned} \tag{2.11}$$

are known as the right Cauchy-Green tensor $\boldsymbol{C}$, and the left Cauchy-Green tensor $\boldsymbol{b}$, present in the reference configuration and in the current configuration, respectively.

2.2.2 Stress tensors and heat flux

In this section, we recall the concept of the internal actions, i.e. contact stresses and heat fluxes, based on Euler's cut principle. Therefore we consider an arbitrary subdomain $\boldsymbol{\Gamma}$ cut out of the body $\mathcal{B}$ as illustrated in Figure 2.2.

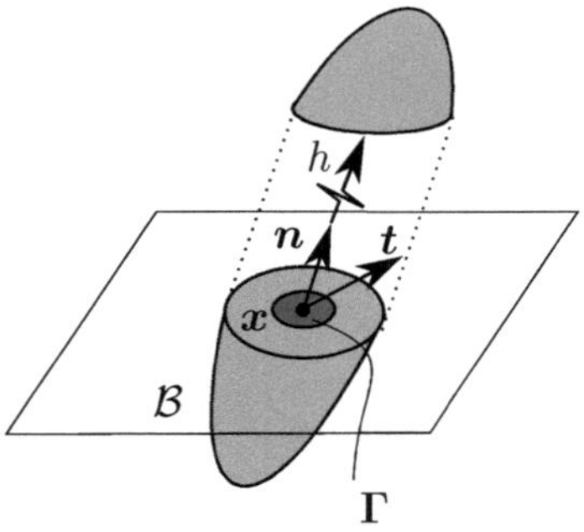

Figure 2.2: Euler's cut principle, with the mechanical traction vector $\boldsymbol{t} = \boldsymbol{\sigma}\boldsymbol{n}$, and the heat flux $h = \boldsymbol{q} \cdot \boldsymbol{n}$

Stress tensors

The Cauchy traction vector $\boldsymbol{t}$, acting within a given spatial point $\boldsymbol{x}$ depends on the orientation of the unit vector $\boldsymbol{n}$. The Cauchy theorem implies that a linear mapping exists

$$\boldsymbol{t} = \boldsymbol{\sigma}\boldsymbol{n}, \tag{2.12}$$

with the Cauchy stress tensor $\boldsymbol{\sigma}$, which can be proved to be symmetrical (by imposing the balance of moments). The Cauchy theorem is also valid for the first Piola-Kirchhoff traction vector $\boldsymbol{T}$, leading to

$$\boldsymbol{T} = \boldsymbol{P}\boldsymbol{N}, \tag{2.13}$$

with the first Piola-Kirchhoff stress tensor $\boldsymbol{P}$ which in spite of the Cauchy stress tensor, is not symmetric. Another stress tensor can be defined through the Cauchy stress tensor $\boldsymbol{\sigma}$ and the Jacobian determinant J

$$\boldsymbol{\tau} = J\boldsymbol{\sigma}, \tag{2.14}$$

in which the tensor $\boldsymbol{\tau}$ is referred as Kirchhoff stress tensor. The second Piola-Kirchoff stress tensor $\boldsymbol{S}$ is defined as the pulled-back of the Kirchoff stress tensor $\boldsymbol{\tau}$. This tensor is a symmetric tensor that does not have a physical interpretation. It is defined as

$$\boldsymbol{S} = \boldsymbol{F}^{-1}\boldsymbol{\tau}\boldsymbol{F}^{-T}. \tag{2.15}$$

The following relationships between the stress tensors defined above can be derived

$$\begin{aligned} \boldsymbol{P} &= J\boldsymbol{\sigma}\boldsymbol{F}^{-T} \\ \boldsymbol{S} &= \boldsymbol{F}^{-1}\boldsymbol{P}, \\ \boldsymbol{S} &= J\boldsymbol{F}^{-1}\boldsymbol{\sigma}\boldsymbol{F}^{-T}. \end{aligned} \tag{2.16}$$

Heat flux

The Stokes heat flux theorem is the thermodynamic equivalent of Cauchy's stress theorem (2.12). It postulates the linear mapping for the heat flux vector h on Γ

$$h = \boldsymbol{q}(\boldsymbol{x}, t) \cdot \boldsymbol{n}, \tag{2.17}$$

with the Cauchy heat flux vector $\boldsymbol{q}$, see Figure 2.2. The Lagrangian counterparts can be defined as

$$H = \boldsymbol{Q} \cdot \boldsymbol{N} \quad \text{with} \quad \boldsymbol{Q} = J\boldsymbol{F}^{-1}\boldsymbol{q}. \tag{2.18}$$

2.2.3 Balance laws

Balance laws are used to refer to the overarching principles or rules of nature. They apply to all types of materials and possess an axiomatic nature. They can be constructed in either a global integral form, affecting the whole body, or a differential local form, affecting each location on the body. Balance rules are used to relate an object's external loads to its internal quantities. If these quantities remain constant during a process, they are referred to as conservation laws. The global and local balance principles of mass, linear momentum, angular momentum, and energy are briefly introduced in this section, as well as the entropy inequality using the first and second laws of thermodynamics.

Balance of mass

With the spatial density field $\rho(\boldsymbol{x}, t)$ and the material density field $\rho_0(\boldsymbol{X})$ the conservation of mass of an infinitesimal volume reads

$$m = \int\limits_{\mathcal{B}_0} \rho_0 \,\mathrm{d}V = \int\limits_{\mathcal{B}} \rho \,\mathrm{d}v = \text{const.} \tag{2.19}$$

Using (2.8) the local form can be deduced as

$$\rho_0 = \rho J. \tag{2.20}$$

Balance of linear momentum

The conservation of linear momentum is a generalization of Newton's second law applied to a continuum body. As such, it is also known as the first Eulerian equation of motion and reads

$$\frac{\mathrm{d}}{\mathrm{d}t} \int\limits_{\mathcal{B}} \boldsymbol{v} \rho \,\mathrm{d}v = \int\limits_{\mathcal{B}} \boldsymbol{b} \,\mathrm{d}v + \int\limits_{\partial\mathcal{B}} \boldsymbol{t} \,\mathrm{d}a \tag{2.21}$$

Here, $\boldsymbol{v}$ is the spatial velocity field, $\boldsymbol{b}$ is an external body force field per unit volume (e.g. gravity). Using the Cauchy stress theorem and the divergence theorem, the local form is given by

$$\nabla \cdot \boldsymbol{P} + \boldsymbol{b} - \rho_0 \dot{\boldsymbol{v}} = \boldsymbol{0} \tag{2.22}$$

Balance of angular momentum

The balance of angular momentum is the rotational analog of the linear momentum and is obtained by differentiating the angular momentum with respect to time

$$\frac{\mathrm{d}}{\mathrm{d}t} \int\limits_{\mathcal{B}} \boldsymbol{x} \times \boldsymbol{v} \rho \,\mathrm{d}v = \int\limits_{\mathcal{B}} \boldsymbol{x} \times \boldsymbol{b} \,\mathrm{d}v + \int\limits_{\partial\mathcal{B}} \boldsymbol{x} \times \boldsymbol{t} \,\mathrm{d}a, \tag{2.23}$$

leading to the local form

$$\boldsymbol{\sigma} = \boldsymbol{\sigma}^T. \tag{2.24}$$

Balance of energy - The first law of thermodynamics

The balance of total energy reads

$$\frac{\mathrm{d}}{\mathrm{d}t}\int\limits_{\mathcal{B}}(\frac{1}{2}\rho\,|\boldsymbol{v}|^2+u)\,\mathrm{d}v=\int\limits_{\partial\mathcal{B}}(\boldsymbol{b}\cdot\boldsymbol{v}+\mathcal{R})\,\mathrm{d}v+\int\limits_{\partial\mathcal{B}}(\boldsymbol{t}\cdot\boldsymbol{v}-h)\,\mathrm{d}a, \tag{2.25}$$

with the stored internal energy per unit volume u and a given heat source $\mathcal{R}$.

The local form

$$\frac{\mathrm{d}}{\mathrm{d}t}u=\boldsymbol{P}:\dot{\boldsymbol{F}}-\nabla\cdot\boldsymbol{Q}+\mathcal{R} \tag{2.26}$$

can be interpreted as the first law of thermodynamics, stating that the rate of change of total energy (kinetic and internal) of a system equals the total energy input into that system, conducted by mechanical and thermal energy.

Dissipation postulate - The second law of thermodynamics

The second law of thermodynamics is founded on the notion of entropy and describes energy changes in terms of their direction. For example, it defines that heat is always transported from a warmer to a cooler location within a continuum. In the subject of irreversible continuum mechanics, such as damage mechanics, entropy is a critical quantity.

The second law of thermodynamics can be defined as

$$\mathcal{T}=\dot{\mathcal{S}}-\mathcal{Q}\geq 0, \tag{2.27}$$

with the total production of entropy $\mathcal{T}$ per unit time, the rate of change of entropy $\dot{\mathcal{S}}$ and the rate of entropy input $\mathcal{Q}$. It can be derived into the local dissipation postulate

$$\mathcal{D}_{\text{int}}=\boldsymbol{P}:\dot{\boldsymbol{F}}-\eta\dot{\theta}-\dot{\Psi}-\frac{1}{\theta}\boldsymbol{Q}\cdot\nabla\theta\geq 0, \tag{2.28}$$

known as the Clausius–Duhem inequality, with the free Helmholtz energy Ψ, the entropy η and the absolute temperature field θ.

2.3 Shell kinematics

This section presents the basic kinematics of Kirchhoff-Love shells used for the formulation of fiber-reinforced sheets in Chapter 5. Shells, defined as planar structural elements with a small thickness relative to their planar dimensions, are used in a variety of engineering applications. Plate and shell theories in continuum mechanics exploit this length scale disparity to reduce the three-dimensional problem of solid mechanics to a two-dimensional one. The two most common formulations are the Mindlin-Reissner and the Kirchhoff-Love models. In the first case, the kinematics is determined by the displacements and rotations of the mean surface, while in the second case only the displacements are needed. With the advent of the isogeometric concept, see Section 2.5, the continuity required by the weak formulation of Kirchhoff-Love shells can be easily achieved.

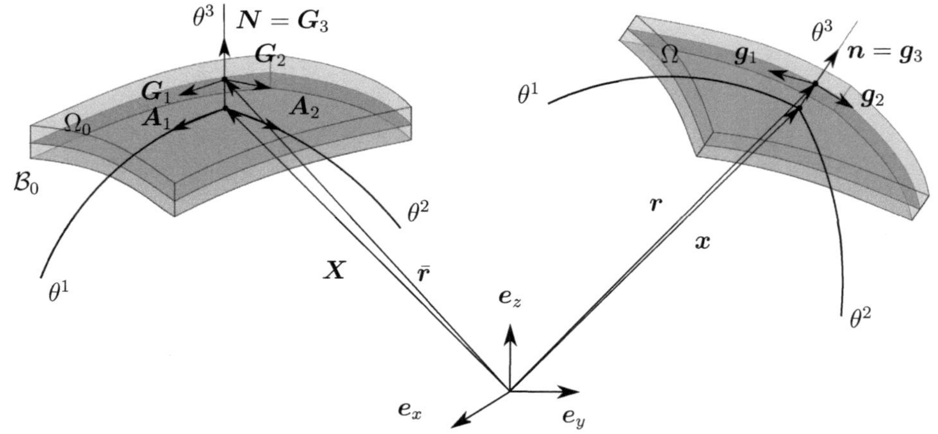

Figure 2.3: Kinematics of the continuum shell in the undeformed and deformed configuration.

To be specific, we consider a three-dimensional continuum shell element as depicted in Figure 2.3, whose points in the Euclidean three-dimensional space $\mathbb{R}^3$ are addressed by the convected coordinates $(\theta^1, \theta^2, \theta^3)$. The reference placement $\boldsymbol{X} : \mathcal{B}_0 \to \mathbb{R}^3$ is the embedding*

$$\boldsymbol{X}(\theta^i) = \bar{\boldsymbol{r}}(\theta^\alpha) + \theta^3 \boldsymbol{N}(\theta^\alpha) \,, \tag{2.29}$$

where $\bar{\boldsymbol{r}} := \bar{\boldsymbol{r}}(\theta^\alpha)$ is the parametrization of the reference mid-surface $\Omega_0 \subset \mathbb{R}^3$, which comes along with the covariant basis $\boldsymbol{A}_\alpha$ and the unit normal vector field $\boldsymbol{N}$ defined by

$$\boldsymbol{A}_\alpha := \bar{\boldsymbol{r}}_{,\alpha} \quad \text{and} \quad \boldsymbol{N} := \frac{\boldsymbol{A}_1 \times \boldsymbol{A}_2}{\|\boldsymbol{A}_1 \times \boldsymbol{A}_2\|} \,, \tag{2.30}$$

where $\bar{\boldsymbol{r}}_{,\alpha} := \frac{\partial \bar{r}}{\partial \theta^\alpha}$. Note that common operations defined on $\mathbb{R}^3$ are extended for vector and tensor fields in a pointwise fashion. For instance, the scalar field $\|\boldsymbol{A}_1 \times \boldsymbol{A}_2\|$ is determined by $\|\boldsymbol{A}_1 \times \boldsymbol{A}_2\|(\theta^\alpha) := \|(\boldsymbol{A}_1(\theta^\alpha) \times \boldsymbol{A}_2(\theta^\alpha)\|$, where we applied the cross product $\times$ of $\mathbb{R}^3$ together with the standard Euclidean norm $\|\boldsymbol{v}\| := \sqrt{(\boldsymbol{v} \cdot \boldsymbol{v})}, \boldsymbol{v} \in \mathbb{R}^3$ induced by the inner product $\cdot$ of $\mathbb{R}^3$. In the reference placement (2.29), the coordinates are chosen such that $\theta^3 \in [-h/2, h/2]$ denotes the coordinate in thickness direction with h as the thickness of the body.

* Greek and Latin indices range in the sets $\{1, 2\}$ and $\{1, 2, 3\}$, respectively. Moreover, we will apply Einstein summation convention for repeated upper and lower indices.

The deformation of the body is described by the function $\boldsymbol{x} : \mathcal{B} \to \mathbb{R}^3$, which incorporates the Kirchhoff hypothesis by the kinematical ansatz

$$\boldsymbol{x}(\theta^i) = \boldsymbol{r}(\theta^\alpha) + \theta^3\, \boldsymbol{n}(\theta^\alpha)\,. \tag{2.31}$$

Here, $\boldsymbol{r} := \boldsymbol{r}(\theta^\alpha)$ is the parametrization of the deformed mid-surface $\Omega \subset \mathbb{R}^3$, which comes along with the covariant basis $\boldsymbol{a}_\alpha$ and the unit normal vector field $\boldsymbol{n}$ defined by

$$\boldsymbol{a}_\alpha := \boldsymbol{r}_{,\alpha} \quad \text{and} \quad \boldsymbol{n} := \frac{\boldsymbol{a}_1 \times \boldsymbol{a}_2}{\|\boldsymbol{a}_1 \times \boldsymbol{a}_2\|}\,. \tag{2.32}$$

The kinematical restriction (2.31) allows to describe the body's deformation exclusively by the geometry of the mid-surface Ω. Therefore, we introduce some important geometric objects, that play a central role in the development of the mechanical theory. For a concise treatment of surface geometry, we refer to [24]. The co- and contravariant metric coefficients of the deformed mid-surface Ω are defined by

$$a_{\alpha\beta} := \boldsymbol{a}_\alpha \cdot \boldsymbol{a}_\beta \quad \text{and} \quad [a^{\alpha\beta}(\theta^\alpha)] := [a_{\alpha\beta}(\theta^\alpha)]^{-1}\,, \tag{2.33}$$

where the square brackets indicate the component matrix. Consequently, it holds that $a_{\alpha\beta}\, a^{\beta\mu} = \delta^\mu_\alpha$. The contravariant basis $\boldsymbol{a}^\alpha$ is introduced by the relation $\boldsymbol{a}^\alpha \cdot \boldsymbol{a}_\beta = \delta^\alpha_\beta$ and can therefore be expressed as $\boldsymbol{a}^\alpha = a^{\alpha\beta}\, \boldsymbol{a}_\beta$, which in turn leads to $a^{\alpha\beta} = \boldsymbol{a}^\alpha \cdot \boldsymbol{a}^\beta$. By definition of the cross-product, we can rewrite $\|\boldsymbol{a}_1 \times \boldsymbol{a}_2\|^2 = [(\boldsymbol{a}_1 \cdot \boldsymbol{a}_1)\,(\boldsymbol{a}_2 \cdot \boldsymbol{a}_2) - (\boldsymbol{a}_1 \cdot \boldsymbol{a}_2)^2] = \det(a_{\alpha\beta})$. The partial derivatives of the covariant base vectors and the normal vectors can be expressed using the Gauss and Weingarten equations

$$\boldsymbol{r}_{,\alpha\beta} = \boldsymbol{a}_{\alpha,\beta} = \Gamma^\sigma_{\alpha\beta}\, \boldsymbol{a}_\sigma + b_{\alpha\beta}\, \boldsymbol{n} \quad \text{and} \quad \boldsymbol{n}_{,\alpha} = -b_{\alpha\beta}\, \boldsymbol{a}^\beta\,, \tag{2.34}$$

where we use the Christoffel symbols of the second kind $\Gamma^\sigma_{\alpha\beta}$ and the covariant components of the curvature tensor $b_{\alpha\beta}$ defined by

$$\Gamma^\sigma_{\alpha\beta} := \boldsymbol{a}^\sigma \cdot \boldsymbol{a}_{\alpha,\beta} \quad \text{and} \quad b_{\alpha\beta} := \boldsymbol{a}_{\alpha,\beta} \cdot \boldsymbol{n}\,. \tag{2.35}$$

Note that $(2.34)_1$ can also be expressed as

$$\boldsymbol{a}_{\alpha,\beta} = \Gamma_{\alpha\beta\sigma}\, \boldsymbol{a}^\sigma + b_{\alpha\beta}\, \boldsymbol{n}\,, \tag{2.36}$$

where

$$\Gamma_{\alpha\beta\sigma} = \boldsymbol{a}_\sigma \cdot \boldsymbol{a}_{\alpha,\beta} = \boldsymbol{a}^\lambda \cdot \boldsymbol{a}_{\alpha,\beta}\, a_{\lambda\sigma} = \Gamma^\lambda_{\alpha\beta}\, a_{\lambda\sigma} \tag{2.37}$$

denotes the Christoffel symbols of the first kind. Analogously, we can define the above introduced objects and relations also for the reference mid-surface Ω_0 using the co- and contravariant metric coefficients $A_{\alpha\beta}$ and $A^{\alpha\beta}$, the contravariant base vectors $\boldsymbol{A}^\alpha$, the Christoffel symbols of the second kind $\bar{\Gamma}^\sigma_{\alpha\beta}$, as well as the covariant components of the curvature tensor $B_{\alpha\beta}$.

The surface-deformation gradient $\boldsymbol{F}$, defined by $\mathrm{d}\boldsymbol{r} = \boldsymbol{F}\,\mathrm{d}\bar{\boldsymbol{r}}$, can be computed by inserting $\mathrm{d}\theta^\alpha = \mathrm{d}\theta^\beta \boldsymbol{A}_\beta \cdot \boldsymbol{A}^\alpha = \bar{\boldsymbol{r}}_{,\beta}\,\mathrm{d}\theta^\beta \cdot \boldsymbol{A}^\alpha = \mathrm{d}\bar{\boldsymbol{r}} \cdot \boldsymbol{A}^\alpha$ into the expression $\mathrm{d}\boldsymbol{r} = \boldsymbol{r}_{,\alpha}\,\mathrm{d}\theta^\alpha = \boldsymbol{a}_\alpha(\mathrm{d}\bar{\boldsymbol{r}} \cdot \boldsymbol{A}^\alpha) = (\boldsymbol{a}_\alpha \otimes \boldsymbol{A}^\alpha)\,\mathrm{d}\bar{\boldsymbol{r}}$ and is consequently given by

$$\boldsymbol{F} = \boldsymbol{a}_\alpha \otimes \boldsymbol{A}^\alpha\,. \tag{2.38}$$

The covariant base vectors $\boldsymbol{g}_i = \boldsymbol{x}_{,i}$ of the three-dimensional body in the deformed configuration are

$$\boldsymbol{g}_\alpha(\theta^i) = \boldsymbol{a}_\alpha(\theta^\alpha) + \theta^3\, \boldsymbol{n}_{,\alpha}(\theta^\alpha) \quad \text{and} \quad \boldsymbol{g}_3(\theta^i) = \boldsymbol{n}(\theta^\alpha) \tag{2.39}$$

and $\boldsymbol{G}_i = \boldsymbol{X}_{,i}$ in the reference configuration

$$\boldsymbol{G}_\alpha(\theta^i) = \boldsymbol{A}_\alpha(\theta^\alpha) + \theta^3\, \boldsymbol{N}_{,\alpha}(\theta^\alpha) \quad \text{and} \quad \boldsymbol{G}_3(\theta^i) = \boldsymbol{N}(\theta^\alpha)\,. \tag{2.40}$$

The covariant metric coefficients $g_{ij} = \boldsymbol{g}_i \cdot \boldsymbol{g}_j$ of the three-dimensional body in the deformed configuration are determined as

$$\begin{aligned} g_{\alpha\beta}(\theta^i) &= a_{\alpha\beta}(\theta^\alpha) - 2\,\theta^3\, b_{\alpha\beta}(\theta^\alpha) + (\theta^3)^2 \boldsymbol{n}_{,\alpha}(\theta^\alpha) \cdot \boldsymbol{n}_{,\beta}(\theta^\alpha)\,, \\ g_{\alpha 3}(\theta^i) &= g_{3\alpha}(\theta^i) = 0\,, \\ g_{33}(\theta^i) &= 1\,, \end{aligned} \tag{2.41}$$

where we applied the definitions (2.33) and $(2.34)_2$ together with the properties $\boldsymbol{n} \cdot \boldsymbol{n} = 1$, $\boldsymbol{a}_\alpha \cdot \boldsymbol{n} = 0$ and $\boldsymbol{n}_{,\alpha} \cdot \boldsymbol{n} = 0$. Analogously, we obtain the covariant metric coefficients $G_{ij} = \boldsymbol{G}_i \cdot \boldsymbol{G}_j$ in the reference configuration

$$\begin{aligned} G_{\alpha\beta}(\theta^i) &= A_{\alpha\beta}(\theta^\alpha) - 2\,\theta^3\, B_{\alpha\beta}(\theta^\alpha) + (\theta^3)^2\, \boldsymbol{N}_{,\alpha}(\theta^\alpha) \cdot \boldsymbol{N}_{,\beta}(\theta^\alpha)\,, \\ G_{\alpha 3}(\theta^i) &= G_{3\alpha}(\theta^i) = 0\,, \\ G_{33}(\theta^i) &= 1\,. \end{aligned} \tag{2.42}$$

Neglecting the quadratic term in equations (2.41) and (2.42) yields

$$g_{\alpha\beta}(\theta^i) = a_{\alpha\beta}(\theta^\alpha) - 2\,\theta^3\, b_{\alpha\beta}(\theta^\alpha) \quad \text{and} \quad G_{\alpha\beta}(\theta^i) = A_{\alpha\beta}(\theta^\alpha) - 2\,\theta^3\, B_{\alpha\beta}(\theta^\alpha) \tag{2.43}$$

assuming a linear behavior through the thickness of the body. Due to the block structure of the metric coefficient matrices, the contravariant base vectors $\boldsymbol{g}^i$ and $\boldsymbol{G}^i$ in the deformed and the reference configuration can be computed as

$$\boldsymbol{g}^\alpha = g^{\alpha\beta}\, \boldsymbol{g}_\beta, \quad \boldsymbol{g}^3 = \boldsymbol{g}_3 = \boldsymbol{n}, \quad [g^{\alpha\beta}(\theta^i)] = [g_{\alpha\beta}(\theta^i)]^{-1}\,, \tag{2.44}$$

and

$$\boldsymbol{G}^\alpha = G^{\alpha\beta}\, \boldsymbol{G}_\beta, \quad \boldsymbol{G}^3 = \boldsymbol{G}_3 = \boldsymbol{N}, \quad [G^{\alpha\beta}(\theta^i)] = [G_{\alpha\beta}(\theta^i)]^{-1}\,, \tag{2.45}$$

respectively. Similar to the surface-deformation gradient $\boldsymbol{F}$, the deformation gradient of the three-dimensional body $\hat{\boldsymbol{F}}$ is

$$\hat{\boldsymbol{F}} = \boldsymbol{g}_i \otimes \boldsymbol{G}^i\,, \tag{2.46}$$

which is easily obtained by $\mathrm{d}\boldsymbol{x} = \boldsymbol{x}_{,i}\, \mathrm{d}\theta^i = \boldsymbol{g}_i(\,\mathrm{d}\boldsymbol{X} \cdot \boldsymbol{G}^i) = (\boldsymbol{g}_i \otimes \boldsymbol{G}^i)\, \mathrm{d}\boldsymbol{X} = \hat{\boldsymbol{F}}\, \mathrm{d}\boldsymbol{X}$.

the right Cauchy-Green deformation tensor for the surface is

$$\boldsymbol{C} := \boldsymbol{F}^T \boldsymbol{F} = C_{\alpha\beta}\, \boldsymbol{A}^\alpha \otimes \boldsymbol{A}^\beta\,, \tag{2.47}$$

where we have defined $C_{\alpha\beta}(a_{\alpha\beta}(\theta^\alpha)) := a_{\alpha\beta}(\theta^\alpha)$. Introducing the difference

$$S^\lambda_{\alpha\beta} := \Gamma^\lambda_{\alpha\beta} - \bar{\Gamma}^\lambda_{\alpha\beta} \quad \text{and} \quad S_{\alpha\beta\lambda} := S^\sigma_{\alpha\beta}\, a_{\sigma\lambda} = \Gamma_{\alpha\beta\lambda} - \bar{\Gamma}^\sigma_{\alpha\beta}\, a_{\sigma\lambda} \,, \tag{2.48}$$

it is convenient to define the second covariant derivative of the deformation with respect to the reference configuration as

$$\begin{aligned} \boldsymbol{r}_{|\alpha\beta} &:= \boldsymbol{r}_{,\alpha\beta} - \bar{\Gamma}^\sigma_{\alpha\beta}\, \boldsymbol{a}_\sigma = (\Gamma^\sigma_{\alpha\beta} - \bar{\Gamma}^\sigma_{\alpha\beta})\, \boldsymbol{a}_\sigma + b_{\alpha\beta}\, \boldsymbol{n} \\ &= S^\sigma_{\alpha\beta}\, \boldsymbol{a}_\sigma + b_{\alpha\beta}\, \boldsymbol{n} = S_{\alpha\beta\sigma}\, \boldsymbol{a}^\sigma + b_{\alpha\beta}\, \boldsymbol{n} \,. \end{aligned} \tag{2.49}$$

Thus, we obtain

$$(\boldsymbol{r}_{|\alpha\beta} - B_{\alpha\beta}\, \boldsymbol{n}) \otimes \boldsymbol{A}^\alpha \otimes \boldsymbol{A}^\beta = \boldsymbol{a}_\sigma \otimes \boldsymbol{S}^\sigma - \boldsymbol{n} \otimes \boldsymbol{\kappa} \tag{2.50}$$

with

$$\boldsymbol{S}^\sigma := S^\sigma_{\alpha\beta}\, \boldsymbol{A}^\alpha \otimes \boldsymbol{A}^\beta \quad \text{and} \quad \boldsymbol{\kappa} := \kappa_{\alpha\beta}\, \boldsymbol{A}^\alpha \otimes \boldsymbol{A}^\beta = (B_{\alpha\beta} - b_{\alpha\beta})\, \boldsymbol{A}^\alpha \otimes \boldsymbol{A}^\beta \,, \tag{2.51}$$

where in-plane and out-of-plane curvatures can be addressed by eight and four components, respectively.

For the matrix material, the right Cauchy-Green tensor reads

$$\hat{\boldsymbol{C}} = \hat{\boldsymbol{F}}^T \hat{\boldsymbol{F}} = g_{ij}\, \boldsymbol{G}^i \otimes \boldsymbol{G}^j \,, \tag{2.52}$$

whose covariant components coincide with the metric components of (2.41); note that higher-order terms in θ^3 are neglected. Due to the dependence of the metric coefficients (2.43) on the first and second fundamental form, it is convenient to introduce the components of the right Cauchy-Green tensor as $\hat{C}_{ij}(a_{\alpha\beta}(\theta^\alpha), \kappa_{\alpha\beta}(\theta^\alpha), \theta^3) = g_{ij}(\theta^i)$. The kinematical restriction in (2.31) does not allow for any change in thickness direction. For a fixed $\theta^3 = \tilde{\theta}^3$, the surface elements of the embedded surfaces in the reference and the deformed configurations $\boldsymbol{X}(\theta^\alpha, \tilde{\theta}^3)$ and $\boldsymbol{x}(\theta^\alpha, \tilde{\theta}^3)$, respectively, are

$$\mathrm{d}a(\theta^\alpha, \tilde{\theta}^3) = \|\boldsymbol{g}_1(\theta^\alpha, \tilde{\theta}^3) \times \boldsymbol{g}_2(\theta^\alpha, \tilde{\theta}^3)\|\, \mathrm{d}\theta^1\, \mathrm{d}\theta^2 = \sqrt{\det(g_{\alpha\beta}(\theta^\alpha, \tilde{\theta}^3)}\, \mathrm{d}\theta^1\, \mathrm{d}\theta^2 \,, \tag{2.53}$$

$$\mathrm{d}A(\theta^\alpha, \tilde{\theta}^3) = \|\boldsymbol{G}_1(\theta^\alpha, \tilde{\theta}^3) \times \boldsymbol{G}_2(\theta^\alpha, \tilde{\theta}^3)\|\, \mathrm{d}\theta^1\, \mathrm{d}\theta^2 = \sqrt{\det(G_{\alpha\beta}(\theta^\alpha, \tilde{\theta}^3))}\, \mathrm{d}\theta^1\, \mathrm{d}\theta^2 \,. \tag{2.54}$$

Consequently, the in-plane Jacobian determinant

$$\hat{J}_0(\theta^i) := \sqrt{\det(\hat{\boldsymbol{C}}(\theta^i))} = \sqrt{\frac{\det(g_{\alpha\beta}(\theta^\alpha, \theta^3))}{\det(G_{\alpha\beta}(\theta^\alpha, \theta^3))}} \tag{2.55}$$

can be interpreted as the areal dilatation, i.e. the ratio between the deformed and the reference area elements $\mathrm{d}a$ and $\mathrm{d}A$, respectively.

2.4 Phase field approach to fracture

The modeling of fracture phenomena with phase-field methods have gained increasing attention in recent years. In this section we will introduce the underlying concepts for the case of non-linear brittle fracture. In subsequent chapters an extended hybrid model will be deduced for porous-ductile fracture in fiber-reinforced thermo-elastic-plastic solids.

2.4.1 Crack regularization

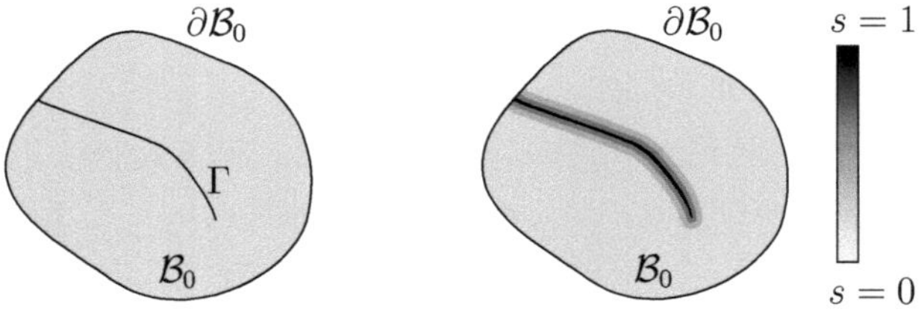

Figure 2.4: A schematic representation of an approximation of an internal discontinuity by a phase-field

We consider an arbitrary elastic body of the bounded domain $\mathcal{B}_0$ in the reference configuration and an evolving internal discontinuity boundary Γ , that represents a set of discrete cracks, as shown in Figure 2.4. In accordance with the classical brittle fracture approach of Griffith [55], crack nucleation and propagation arises upon the attainment of a critical local fracture energy density g_c. The total potential energy of the body then reads

$$W^\mathrm{pot} = W^\mathrm{e} + W^\mathrm{f} = \int_{\mathcal{B}_0} \Psi_\mathrm{e} \,\mathrm{d}V + \int_{\Gamma} g_\mathrm{c} \,\mathrm{d}A, \tag{2.56}$$

whose minimization can describe the entire fracture process, controlled by a potential elastic strain energy term and a term related to the work required for the creation of new surfaces.

The tracking of the crack discontinuity boundaries in classical fracture approaches often requires complex and costly computations, particularly when considering complex shaped cracks or even crack branching. The underlying concept of the phase-field formulation is to approximate the discrete fracture surface with a continuous scalar field, the so called phase-field $\mathfrak{s}(\boldsymbol{X}, t)$, where the value $\mathfrak{s} = 0$ refers to the undamaged and $\mathfrak{s} = 1$ to the fully broken state of the material. Thus, the crack surface energy can be approximated as

$$\int_{\Gamma} g_\mathrm{c} \,\mathrm{d}A \approx \int_{\mathcal{B}_0} \Psi_\mathrm{f}(\mathfrak{s}) \,\mathrm{d}V = \int_{\mathcal{B}_0} g_\mathrm{c} \,\gamma(\mathfrak{s}) \,\mathrm{d}V, \tag{2.57}$$

with a crack density functional γ, typically defined as

$$\gamma(\mathfrak{s}, \nabla(\mathfrak{s})) = \frac{1}{2\,l}\,\mathfrak{s}^2 + \frac{l}{2}\,\nabla(\mathfrak{s}) \cdot \nabla(\mathfrak{s}). \tag{2.58}$$

The smooth bases of isogeometric analysis presented in Section 2.5.2 provide a convenient computational framework to apply higher-order regularization and thus obtain better accuracy and convergence rates, see Borden et al. [19]. The common fourth-order approach reads

$$\gamma(\mathfrak{s}, \nabla(\mathfrak{s}), \Delta(\mathfrak{s})) = \frac{1}{4\,l}\,\mathfrak{s}^2 + \frac{l}{2}\,\nabla(\mathfrak{s}) \cdot \nabla(\mathfrak{s}) + \frac{l^3}{4}\,\Delta(\mathfrak{s})\,\Delta(\mathfrak{s}). \tag{2.59}$$

The length scale parameter l in the the crack density functionals characterizes the width of the smeared crack profile. For $\mathfrak{s} \to 0$, the approximation converges to the sharp crack Γ. It can also be considered as a material property governing the critical value of stress required for the crack initiation or growth. For a one-dimensional bar under tension, l can be derived with the critical tensile stress $\sigma_{\rm cr}$ as

$$l = \frac{27\, E\, g_{\rm c}}{256\, {\sigma_{\rm cr}}^2}, \tag{2.60}$$

see [13] for details. Though an extension in higher dimensions and complex mixed-mode fracture analysis is difficult, this can give a rough estimation on how to choose l. In order to correctly resolve the steep gradient of the smooth crack profile, a fine element discretization should be employed in the region where a crack is expected.

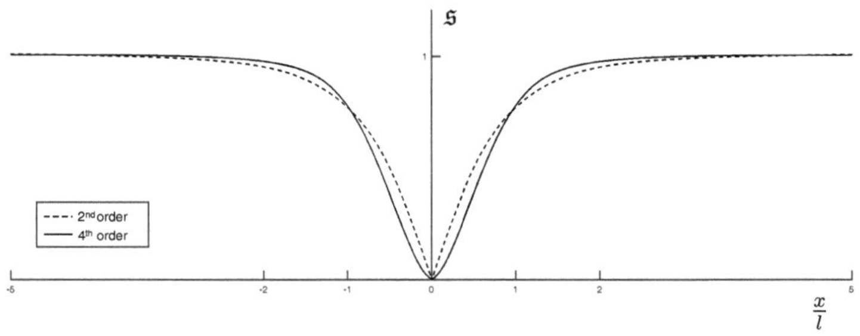

Figure 2.5: Analytical solution for unidimensional phase-field profiles for second- and fourth-order formulations

The unidimensional analytical solution of the corresponding Euler equations of the crack density functionals (2.58) and (2.59), after minimizing $\int_{\mathcal{B}_0} g_{\rm c}\, \gamma(\mathfrak{s})\, {\rm d}V$ and imposing the initial condition $\mathfrak{s}(0) = 0$, are depicted in Figure 2.5. Note that the higher-order formulation has a slightly narrower profile, meaning that a smaller region needs to have a fine mesh in order to correctly resolve the crack. For this and for the abovementioned reasons, the fourth-order model is adopted for the simulations in this work.

2.4.2 Elastic strain energy

Next, the loss of material stiffness in the crack is incorporated via the degradation of the strain energy through the phase-field. The simplest constitutive description would be

$$\Psi^e(\boldsymbol{F}, \mathfrak{s}) = g(\mathfrak{s})\, \Psi(\boldsymbol{F}) \tag{2.61}$$

with the linear degradation function $g(\mathfrak{s}) = (1 - \mathfrak{s})$. Assuming that fracture requires a local state of tension, a physically more reasonable formulation distinguishes between crack-insensitive compressive ($^-$) and crack-sensitive tensile components ($^+$). A common additive decomposition of the strain energy density reads

$$\Psi^e(\boldsymbol{F}, \mathfrak{s}) = g(\mathfrak{s})\, \Psi^+(\boldsymbol{F}) + \Psi^-(\boldsymbol{F}), \tag{2.62}$$

where the degradation induced by the phase-field acts only on the tension term of the strain energy function. A possible formulation for large deformation, proposed by Amor et al. [8] and Borden [18], is based on a the following decomposition of the strain energy into volumetric and deviatoric parts

$$\begin{aligned} &\text{if} \quad J < 1 \begin{cases} \Psi^+ = \Psi_{\text{dev}}(\boldsymbol{F}) \\ \Psi^- = \Psi_{\text{vol}}(J) \end{cases} \\ &\text{elseif} \quad J \geq 1 \begin{cases} \Psi^+ = \Psi_{\text{dev}}(\boldsymbol{F}) + \Psi_{\text{vol}}(J) \\ \Psi^- = 0 \end{cases} \end{aligned} \quad . \tag{2.63}$$

Multiplicative decomposition of crack-sensitive and -insensitive components have also been proposed over the principal stretches λ_a for the non-linear regime. Hesch et al. [126] introduced an anisotropic split via $\lambda_a = \lambda_a^- \lambda_a^+$ with

$$\lambda_a^\pm = \frac{(\lambda_a - 1) \pm \mid \lambda_a - 1 \mid}{2} + 1 \tag{2.64}$$

leading to the fracture insensitive part of the deformation gradient

$$\tilde{\boldsymbol{F}} = \sum_a (\lambda_a^+)^{g(\mathfrak{s})} \lambda_a^- \boldsymbol{n}_a \otimes \boldsymbol{N}_a, \tag{2.65}$$

where $\boldsymbol{n}_a$ and $\boldsymbol{N}_a$ denote the principal directions of the left and right stretch tensors, respectively. Since the just mentioned decomposition is not compatible with a standard elastoplasticity formulation, which relies on different mechanisms for the deviatoric and volumetric contributions, we introduce a novel multiplicative decomposition in this work, defining the fracture insensitive, isochoric part of the deformation gradient and the fracture insensitive of the Jacobian determinant as

$$\tilde{\bar{\boldsymbol{F}}} = \sum_a (J^{-1/d} \lambda_a)^{g(\mathfrak{s})} \boldsymbol{n}_a \otimes \boldsymbol{N}_a \quad \text{and} \quad \tilde{J} = \begin{cases} J^{g(\mathfrak{s})} & \text{if} \quad J > 1 \\ J & \text{else} \end{cases} . \tag{2.66}$$

The associated elastic strain energy reads

$$\Psi^e(\boldsymbol{F}, \mathfrak{s}) = \Psi_{\text{dev}}(\tilde{\bar{\boldsymbol{F}}}(\boldsymbol{F}, \mathfrak{s})) + \Psi_{\text{vol}}(\tilde{J}(\boldsymbol{F}, \mathfrak{s})) \tag{2.67}$$

for which typical decoupled strain energies can be taken into account.

Note that more alternative formulations for the tension-compression split have also been presented, like decompositions based on the symmetric and antisymmetric parts of the strain tensor [51] or with respect to the crack orientation [119].

The degradation function $g(\mathfrak{s})$ controls the transition of the behavior of the material from the intact to the cracked state and has generally the following properties

$$g(0) = 1, \quad g(1) = 0 \quad \text{and} \quad g'(1) = 0. \tag{2.68}$$

Commonly, the following quadratic degradation function is considered

$$g(\mathfrak{s}) = (1 - \mathfrak{s})^2 + \varepsilon. \tag{2.69}$$

The positive small factor $\varepsilon \approx 0$ can be used for avoiding zero stiffness of the material in a fully cracked state. With a more general cubic degradation function in the form

$$g(\mathfrak{s}) = (a_{\mathrm{g}} - 2)\,(1 - \mathfrak{s})^3 + (3 - a_{\mathrm{g}})\,(1 - \mathfrak{s})^2 + \varepsilon \tag{2.70}$$

a small but positive slope of the function at $\mathfrak{s} = 0$ allowing better crack nucleation is achieved with small positive values for the degradation modeling parameter a_{g}. For brittle fracture, a perfectly linear stress-strain relationship, i.e. a pure elastic behavior with no softening prior to crack nucleation can be observed with with $a_{\mathrm{g}} \to 0$, see [18].

2.4.3 Balance equation of the phase-field

By varying the density functions of the stored energy with respect to $\mathfrak{s}$, we obtain the additional constitutive quantities

$$\begin{aligned} \mathcal{H} &= -\delta_{\mathfrak{s}} \widehat{\Psi}^e = -\partial_{\mathfrak{s}} \widehat{\Psi}^e \\ r^f &= \delta_{\mathfrak{s}} \widehat{\Psi}^f \quad = \partial_{\mathfrak{s}} \widehat{\Psi}^f - \mathrm{Div}[\partial_{\nabla \mathfrak{s}} \widehat{\Psi}^f] = \frac{g_c}{l_f}\, \mathfrak{s} - g_c\, l_f\, \Delta \mathfrak{s}, \end{aligned}$$

namely the driving force of the phase-field $\mathcal{H}$ and the crack resistance force r^f. Note that the second order functional Γ is considered here. The strong form of the mechanical system can then be extended with

$$\mathcal{H} = \frac{g_c}{l_f}\, \mathfrak{s} - g_c\, l_f\, \Delta \mathfrak{s}. \tag{2.71}$$

2.5 Isogeometric analysis

In this section we provide an introduction of the isogeometric concept used for all geometrical discretizations in this work. In conventional finite element analysis, the spline-based CAD geometry is approximated with low-order, usually linear, Lagrange polynomials basis functions. The term "isogeometric analysis" implies that the FEA model employs the same spline-based description as in the CAD geometry. Additionally, it provides higher continuity of the finite element approximation, needed for certain models, i.e. the fourth-order phase-field model, the Kirchhoff-Love theory and the strain-gradient formulation of the fiber material.

2.5.1 B-splines and NURBS

Univariate B-splines

Univariate B-spline functions are piecewise polynomial functions with compact support. They are defined in parametric space using a so-called knot vector denoted Ξ . In a one-dimensional space the knot vector is a set of finite and monotonically increasing sequence of real numbers $\Xi = [\xi_1, \xi_2, \ldots, \xi_{\mathrm{n}+p+1}]$, where p is the polynomial degree and n is the number of basis functions of the B-spline.

Using the convention that a fraction in front of the basis functions is set equal to zero in the case of the denominator being zero (i.e. $\frac{0}{0} := 0$), the B-spline functions are defined recursively on the vector of knots with the Cox-de Boor formula [27] as

$$B_p^i(\xi) = \frac{\xi - \xi_i}{\xi_{i+p} - \xi_i} B_{p-1}^i(\xi) + \frac{\xi_{i+p+1} - \xi}{\xi_{i+p+1} - \xi_{i+1}} B_{p-1}^{i+1}(\xi), \tag{2.72}$$

beginning with

$$B_0^i(\xi) = \begin{cases} 1 & \text{if } \xi_i \leq \xi < \xi_{i+1} \\ 0 & \text{otherwise} \end{cases} \tag{2.73}$$

for $p = 0$. It can be noted that for $p = 0$ and $p = 1$, the basis functions of isogeometric analysis are identical to those of the standard piecewise constant and linear finite elements, respectively.

Each knot represents a coordinate value in the parametric space. If the knot vector is chosen equal to a set of following integers, we refer to natural B-spline. If the knots are distributed uniformly, the vector of knot is said to be uniform. It is said to be open if the first and last knots are repeated $p + 1$times. Open knot vectors are standard in the CAD literature. In one dimension, basis functions formed from open knot vectors are interpolatory at the ends of the parametric space interval, $[\xi_1, \xi_{\mathrm{m}+p+1}]$, and at the corners of patches in multiple dimension but they are not, in general, interpolatory at interior knots. An example of quadratic basis functions for an open, uniform knot vector is presented in Figure 2.6.

In general, basis functions of degree p have $p - 1$ continuous derivatives. If a knot is repeated k times (with $k < p$), then the number of continuous derivatives decreases by k. When the multiplicity of a knot is exactly p (i.e. $k = p$), the B-spline basis functions are C^0 continuous at that knot point. This is illustrated in Figure 2.6 at the location of the repeated knot $\xi = 4$. Furthermore, if the multiplicity of the knot point is $p + 1$ (i.e. $k = p + 1$) the B-spline basis functions are discontinuous at that knot point and as a result two separate B-splines patches are formed.

In the perspective of finite element analysis, the univariate B-spline basis functions, as well as the multivariate B-spline and NURBS basis functions have important features, namely

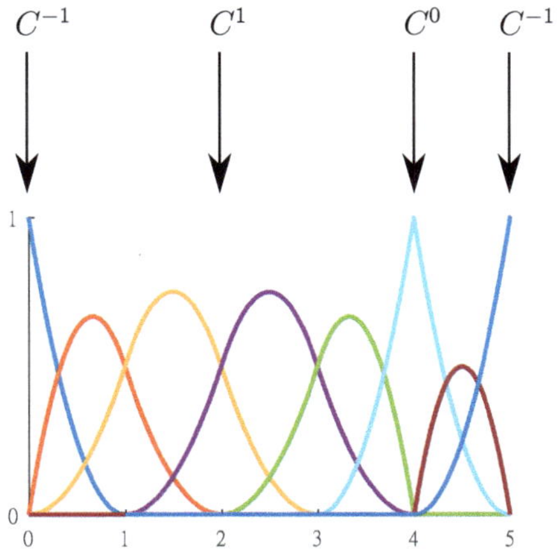

Figure 2.6: Quadratic basis functions for the open-uniform knot vector $\xi = [0, 0, 0, 1, 2, 3, 4, 4, 5, 5, 5]$ with varying degrees of regularity.

- Partition of unity $\sum_i B_p^i(\xi) \equiv 1.$
- Nonnegativity over the entire domain $B_p^i(\xi) \geq 0$
- Compact support of each $B_p^i(\xi)$ in $[\xi_i, \xi_{i+p+1}]$,
- Linear independence $\sum_i B_p^i(\xi) c_i \equiv 0 \quad \Leftrightarrow \quad c_i = 0$

The derivatives of B-spline functions are represented in terms of lower-order B-spline basis. Thus, the first derivative of the ith B-spline basis function of degree p is given by

$$\frac{\mathrm{d}}{\mathrm{d}\xi} B_p^i(\xi) = \frac{p}{\xi_{i+p} - \xi_i} B_{p-1}^i(\xi) - \frac{p}{\xi_{i+p+1} - \xi_{i+1}} B_{p-1}^{i+1}(\xi) \tag{2.74}$$

and consequently the jth derivative with

$$\frac{\mathrm{d}^j}{\mathrm{d}\xi^k} B_p^i(\xi) = \frac{p}{\xi_{i+p} - \xi_i} \left(\frac{\mathrm{d}^{j-1}}{\mathrm{d}\xi^{j-1}} B_{p-1}^i(\xi) \right) - \frac{p}{\xi_{i+p+1} - \xi_{i+1}} \left(\frac{\mathrm{d}^{j-1}}{\mathrm{d}\xi^{j-1}} B_{p-1}^{i+1}(\xi) \right). \tag{2.75}$$

Multivariate B-splines

One can easily extend the previous concepts to define multivariate B-splines for dimension $d \in \{2, 3\}$ by using the dyadic product $\boldsymbol{\Xi} = \Xi_1 \otimes \ldots \otimes \Xi_d$ of univariate knot vectors Ξ_i, with the direction in the parameter space $i \in \{1, \ldots, d\}$. More precisely multivariate B-splines are defined as

$$B^G = B_p^i(\boldsymbol{\xi}) = \prod_{l=1}^{d} B_{p_l}^{i_l}(\xi^l), \tag{2.76}$$

with the associated univariate B-spline basis functions $B^{i_l}_{p_l}(\xi^l)$ where the multi indexes $\boldsymbol{i} = [i_1, \ldots, i_d]$ and $\boldsymbol{p} = [p_1, \ldots, p_d]$ represent the position in the tensor product and the polynomial degree in the parameter space, respectively. The assignment of the global numbering index G to the multi index $\boldsymbol{i}$ can be found [26].

Using (2.74) and (2.75), the jth partial derivative of the Gth multivariate B-spline with respect to ξ^l can be obtained as

$$\frac{\partial^k B^G}{\partial \xi^{k_l}} = \frac{\partial^k}{\partial \xi^{k_l}} B^{i_l}_{p_l}(\xi^l) \prod_{\substack{\hat{l}=1 \\ \hat{l}\neq l}}^{d} B^{i_{\hat{l}}}_{p_{\hat{l}}}(\xi^{\hat{l}}), \tag{2.77}$$

NURBS

B-splines are convenient for free-form modeling, but they lack the ability to exactly represent some simple shapes such as circles, cylinders and ellipsoids. This limitations can be overcome with the extension to non-uniform rational B-splines (NURBS), which enable more control over local geometry shape due to non-uniform treatment of each control point influence on the curve. This is obtained by the use of weights $w_{\boldsymbol{i}}$ for each control point respectively according to

$$N^G = N^{\boldsymbol{i}}_{\boldsymbol{p}}(\boldsymbol{\xi}) = \frac{\prod\limits_{l=1}^{d} B^{i_l}_{p_l}(\xi^l)\, w_{\boldsymbol{i}}}{\sum\limits_{\boldsymbol{j}} \prod\limits_{l=1}^{d} B^{j_l}_{p_l}(\xi^l)\, w_{\boldsymbol{j}}}, \tag{2.78}$$

Note that by setting the same value for each weight we obtain the original B-spline.

By applying the quotient rule, the partial derivative of N^A with respect to ξ^l reads

$$\frac{\partial N^G}{\partial \xi^l} = \frac{\sum\limits_{B} B^B w_B \frac{\partial B^A}{\partial \xi^l} - B^A \sum\limits_{B} \frac{\partial B^B}{\partial \xi^l} w_B}{\left(\sum\limits_{B} B^B w_B\right)^2} w_A. \tag{2.79}$$

Higher jth order derivatives can be obtained by

$$\frac{\partial^k N^G}{\partial \xi^{l^j}} = \frac{\frac{\partial^j B^G}{\partial \xi^{l^j}} w_G - \sum\limits_{m=1}^{k} \binom{k}{m} \frac{\partial^{k-m} N^G}{\partial \xi^{l^{k-m}}} \sum\limits_{B} \frac{\partial^m B^B}{\partial \xi^{l^m}} w_B}{\sum\limits_{B} B^B w_B}. \tag{2.80}$$

Hierarchical refinement

The univariate and multivariate B-splines and NURBS can be refined by different ways without changing the shape of the geometry.

Through **knot insertion**, the solution space is enriched by adding more basis functions of the same order while keeping the spline unchanged. Insertion of new knot values clearly has similarities with the classical h-refinement strategy in FEA.

In the process of **order elevation**, the B-Spline/NURBS-order is raised by increasing the multiplicity of each knot (and without adding any new knots values). It can be thought of as p-refinement in FEA.

Lastly, **hierarchical refinement** of multivariate B-spline and NURBS bases have been proposed in [35] with preservation of the global properties in terms of smoothness and continuity of the unrefined mesh.

2.5.2 Geometric discretization

In the isogeometric analysis framework, the computational domain is not approximated, but exactly described. To be specific, we introduce a non-linear transformation from the parametric space $\hat{\mathcal{B}}$ into the physical space $\mathcal{B}$

$$\begin{aligned} \mathcal{T}: \quad & \hat{\mathcal{B}} \rightarrow \mathcal{B} \\ & \boldsymbol{\xi} \rightarrow \boldsymbol{X}. \end{aligned} \tag{2.81}$$

By associating the NURBS basis functions N^G with a net of control points $\boldsymbol{X}_G \in \mathbb{R}^d$ related to the geometry under consideration, i.e. a curve or surface as seen in Figure 2.7, we obtain

$$\mathcal{G} = \mathcal{T}(\boldsymbol{\xi}) = \sum_G N^G \boldsymbol{X}_G = \sum_i N_p^i(\boldsymbol{\xi}) \boldsymbol{X}_i. \tag{2.82}$$

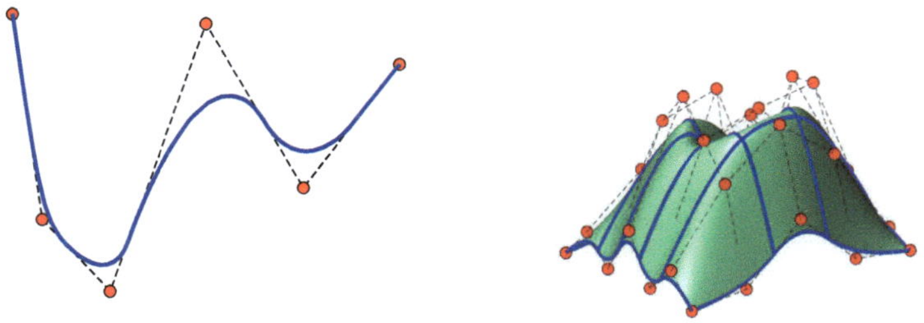

Figure 2.7: Physical mesh and control mesh of a quadratic NURBS curve (left) and a quadratic NURBS surface (right). The red points denote the control points.

Accordingly, rational approximations of the deformed geometry $\boldsymbol{\varphi}$ and its variation $\delta\boldsymbol{\varphi}$ are defined as

$$\boldsymbol{\varphi}^{\mathrm{h}} = \sum_{G} N^{G} \boldsymbol{X}_{G}, \quad \delta\boldsymbol{\varphi}^{\mathrm{h}} = \sum_{G} N^{G} \delta\boldsymbol{X}_{G}, \tag{2.83}$$

which we can insert into the corresponding terms of the balance equations to obtain the discrete formulation of our problem.

3 Phase-Field Modeling of Non-Linear Thermo-Porous Ductile Fracture

In this chapter, we present a framework for simulating porous-ductile fracture in isotropic thermo-elastic-plastic solids subjected to large deformations. The proposed model is based on a triple multiplicative decomposition of the deformation gradient and an exponential update scheme for the plastic return map in the time discrete setting. In addition, a modified Gurson–Tvergaard–Needleman GTN-type strain-gradient thermoplasticity model is combined with a second-gradient phase-field fracture approach to account for a temperature-dependent growth of voids on microscale followed by crack initiation and propagation on macroscale. The multi-physical formulation is completed by incorporating an energy transfer into the thermal field such that the temperature distribution depends on the evolution of the plastic strain and the crack phase-field. Eventually, this physically comprehensive fracture formulation is validated by experimental data and applied to complex three-dimensional geometries. Based on the available comparative experimental data, the specific material models and parameters in this chapter are provisionally adapted for steel-like materials. Generally applicable to thermo-elastic-plastic solids, the framework and formulations presented here are specifically adapted to the final polymeric matrix material in the subsequent chapters.

3.1 Configuration and kinematics

As introduced in Chapter 2, we consider a body of interest $\mathcal{B}_0 \in \mathbb{R}^3$ in its reference configuration, along with the deformation mapping $\boldsymbol{\varphi}(\boldsymbol{X},t)$, the material deformation gradient $\boldsymbol{F}(\boldsymbol{X},t)$, the crack phase-field $\mathfrak{s}(\boldsymbol{X},t)$ and the absolute temperature $\theta(\boldsymbol{X},t)$.

Regarding the plasticity, we introduce the equivalent plastic strain $\alpha(\boldsymbol{X},t)$ and its dual dissipative resistance force $r^{\mathrm{p}}(\boldsymbol{X},t)$ as

$$\alpha(\boldsymbol{X},t) : \mathcal{B}_0 \times \mathcal{T} \to \mathbb{R} \quad \text{and} \quad r^{\mathrm{p}}(\boldsymbol{X},t) : \mathcal{B}_0 \times \mathcal{T} \to \mathbb{R}. \tag{3.1}$$

The plastic deformation gradient $\boldsymbol{F}^{\mathrm{p}}$ represents an additional strain measure field

$$\boldsymbol{F}^{\mathrm{p}}(\boldsymbol{X},t) : \mathcal{B}_0 \times \mathcal{T} \to \mathbb{R}^{3\times 3}, \quad \text{with} \quad J^{\mathrm{p}} = \det[\boldsymbol{F}^{\mathrm{p}}] \geq 1. \tag{3.2}$$

Within the framework of a GTN-type plasticity model, the void volume fraction f is introduced as a micromechanically motivated damage parameter for growing cavitation in plastic solids

$$f(\boldsymbol{X},t) : \mathcal{B}_0 \times \mathcal{T} \to [f_0, f_{\mathrm{f}}] \quad \text{with} \quad f = \frac{\text{volume of voids}}{\text{total volume}}, \tag{3.3}$$

where f_0 is the initial porosity and f_{f} is the final void volume fraction. The above introduced variables characterize a multi-field setting of porous-ductile fracture in thermo-elastic-plastic solids based on five independent fields

$$\mathfrak{U} := \{\boldsymbol{\varphi}, \mathfrak{s}, \theta, \alpha, r^{\mathrm{p}}\}, \tag{3.4}$$

the finite deformation map $\boldsymbol{\varphi}$, the absolute temperature θ, the crack phase-field $\mathfrak{s}$, the equivalent plastic strain α and the dissipative plastic resistance force r^{p}. Note that the Lagrangian plastic deformation map $\boldsymbol{F}^{\mathrm{p}}$ and the void volume fraction f will be condensed within the balance equations.

As is common in nonlinear elastoplasticity, we apply a multiplicative split of the deformation gradient and its determinant, and obtain the elastic components

$$\boldsymbol{F}^{\mathrm{e}} = \boldsymbol{F}\,(\boldsymbol{F}^{\mathrm{p}})^{-1} \quad \text{and} \quad J^{\mathrm{e}} = J\,(J^{\mathrm{p}})^{-1}. \tag{3.5}$$

Postulating that fracture requires a local state of tensile/shear deformation as discussed in [8, 88, 37, 40], we define the fracture insensitive, isochoric part of the deformation gradient and the fracture insensitive of the Jacobian determinant as follows

$$\tilde{\bar{\boldsymbol{F}}}^{\mathrm{e}} = \sum_a \tilde{\bar{\lambda}}^{\mathrm{e}}_a\, \boldsymbol{n}_a \otimes \boldsymbol{N}_a \quad \text{and} \quad \tilde{J}^{\mathrm{e}} = \begin{cases} (J^{\mathrm{e}})^{g(\mathfrak{s})} & \text{if} \quad J^{\mathrm{e}} > 1 \\ J^{\mathrm{e}} & \text{else} \end{cases}. \tag{3.6}$$

The degradation function $g(\mathfrak{s})$ is often defined by the following cubic function $g(\mathfrak{s}) = (a_{\mathrm{g}} - 2)\,(1-\mathfrak{s})^3 + (3 - a_{\mathrm{g}})\,(1-\mathfrak{s})^2$, where $a_{\mathrm{g}} \in (0, 2]$ is a degradation modeling parameter. Note that we use a simplified quadratic degradation function $(1 - \mathfrak{s})^2$, which is obtained by setting $a_{\mathrm{g}} = 2$. Moreover, $\tilde{\bar{\lambda}}^{\mathrm{e}}_a = (\bar{\lambda}^{\mathrm{e}}_a)^{g(\mathfrak{s})}$ are distortional stretches, defined via the isochoric, elastic parts of the principal stretches $\bar{\lambda}^{\mathrm{e}}_a = (J^{\mathrm{e}})^{-1/d}\lambda^{\mathrm{e}}_a$, and finally, the vectors $\boldsymbol{n}_a$ and $\boldsymbol{N}_a$ represent the principal directions of the left and right stretch tensors, see [37].

Furthermore, in order to subsequently derive objective (frame-independent) constitutive relations, we introduce the elastic left Cauchy-Green

$$\boldsymbol{b}^{\mathrm{e}} = \boldsymbol{F}\,(\boldsymbol{C}^{\mathrm{p}})^{-1}\,\boldsymbol{F}^{\mathrm{T}} \tag{3.7}$$

with $\boldsymbol{C}^{\mathrm{p}} = (\boldsymbol{F}^{\mathrm{p}})^{\mathrm{T}}\,\boldsymbol{F}^{\mathrm{p}}$. Consequently, the fracture insensitive part of the elastic left Cauchy-Green tensor reads

$$\tilde{\bar{\boldsymbol{b}}}^{\mathrm{e}} = \tilde{\bar{\boldsymbol{F}}}^{\mathrm{e}}\,(\tilde{\bar{\boldsymbol{F}}}^{\mathrm{e}})^{\mathrm{T}} = \sum_a (\tilde{\bar{\lambda}}^{\mathrm{e}}_a)^2\,\boldsymbol{n}_a \otimes \boldsymbol{n}_a. \tag{3.8}$$

In contrast to the isochoric von Mises plasticity using $J = \sqrt{\det[\boldsymbol{b}^{\mathrm{e}}]}$, the volumetric plastic void growth considered in this work leads to the total determinant

$$J = \sqrt{\det[\boldsymbol{b}^{\mathrm{e}}]}\,\sqrt{\det[\boldsymbol{C}^{\mathrm{p}}]}. \tag{3.9}$$

3.2 Variational formulation

Next, we propose the constitutive framework for the thermomechanical damage model. To be specific, we introduce constitutive energetic and dissipative functions and derive the necessary relations and evolution laws to formulate the multi-field variational problem.

3.2.1 Energetic response function

By coupling gradient thermoplasticity with gradient damage mechanics, the density function of the stored energy $\widehat{\Psi}$ for the multi-field problem is defined by

$$\widehat{\Psi} = \widehat{\Psi}^{\mathrm{e}}(\boldsymbol{F}, \boldsymbol{F}^{\mathrm{p}}, \mathfrak{s}, \theta) + \widehat{\Psi}^{\theta}(\theta) + \widehat{\Psi}^{\mathrm{p}}(\alpha, \nabla\alpha, \theta) + \widehat{\Psi}^{\mathrm{f}}(\mathfrak{s}, \nabla\mathfrak{s}, \Delta\mathfrak{s}). \tag{3.10}$$

The elastic contribution is composed of volumetric and deviatoric parts as

$$\widehat{\Psi}^{\mathrm{e}} = \widehat{\Psi}^{\mathrm{e}}_{\mathrm{vol}}(\tilde{J}^{\mathrm{e}}(J, J^{\mathrm{p}}, \mathfrak{s}), \theta) + \widehat{\Psi}^{\mathrm{e}}_{\mathrm{dev}}(\bar{\tilde{\boldsymbol{F}}}^{\mathrm{e}}(\boldsymbol{F}, \boldsymbol{F}^{\mathrm{p}}, J, \mathfrak{s})) \tag{3.11}$$

where, without prejudice to the generality, we consider in the examples an extended Neo-Hookean material law given as follows

$$\widehat{\Psi}^{\mathrm{e}}_{\mathrm{vol}}(\tilde{J}^{\mathrm{e}}, \theta) = \frac{\kappa}{2}\left(\frac{(\tilde{J}^{\mathrm{e}})^2 - 1}{2} - \ln[\tilde{J}^{\mathrm{e}}]\right) - \frac{3}{2}\,\beta\,\kappa\,(\theta - \theta_0)\left(\tilde{J}^{\mathrm{e}} - \frac{1}{\tilde{J}^{\mathrm{e}}}\right) \tag{3.12}$$

and

$$\widehat{\Psi}^{\mathrm{e}}_{\mathrm{dev}}(\bar{\tilde{\boldsymbol{F}}}^{\mathrm{e}}) = \frac{\mu}{2}\,(\bar{\tilde{\boldsymbol{F}}}^{\mathrm{e}} : \bar{\tilde{\boldsymbol{F}}}^{\mathrm{e}} - 3). \tag{3.13}$$

Therein, $\mu > 0$ and $\kappa > 0$ denote the shear modulus and the bulk modulus, respectively. θ_0 is a reference temperature and β is the linear thermal expansion coefficient. Moreover, for a compact representation we introduce the abbreviation $\bar{\tilde{\boldsymbol{F}}}^{\mathrm{e}} : \bar{\tilde{\boldsymbol{F}}}^{\mathrm{e}} = \sum_a (\bar{\tilde{\lambda}}^{\mathrm{e}}_a)^2$. Next, the purely thermal contribution to the energy (3.10) is assumed in the simple form

$$\widehat{\Psi}^{\theta}(\theta) = c\left(\theta - \theta_0 - \theta\ln\left(\frac{\theta}{\theta_0}\right)\right), \tag{3.14}$$

where $c \geq 0$ is a constant parameter representing the specific heat capacity. Satisfying the heat conduction inequality via Duhamel's law of heat conduction, we define the Piola-Kirchhoff heat flux vector as

$$\boldsymbol{Q}(\boldsymbol{F}, \mathfrak{s}, \theta, \nabla\theta) := -\boldsymbol{K}(\boldsymbol{F}, \mathfrak{s}, \theta)\nabla\theta. \tag{3.15}$$

Phenomenologically motivated, we formulate the material thermal conductivity tensor $\boldsymbol{K}$ in terms of the phase-field parameter $\mathfrak{s}$

$$\boldsymbol{K}(\boldsymbol{F}, \mathfrak{s}, \theta) := \left[K_0\,(1 - w_{\mathrm{K}}\,(\theta - \theta_0))\,(1 - \mathfrak{s}) + K^{\mathrm{conv}}\,\mathfrak{s}\right]\boldsymbol{C}^{-1}, \tag{3.16}$$

so that when a fracture occurs, the conduction degenerates locally and leads to a pure convection problem where the heat transfer depends on the crack opening width. Moreover, w_{K} is a thermal softening parameter, K_0 is a conductivity parameter related to the reference temperature, K^{conv} is a convection parameter and $\boldsymbol{C} = \boldsymbol{F}^{\mathrm{T}}\boldsymbol{F}$ denotes the right Cauchy-Green tensor.

Based on the particular formulation for gradient plasticity in [2, 5, 37, 35], the plastic energy contribution is defined in terms of the equivalent plastic strain α, its gradient $\nabla\alpha$ and the temperature field θ as

$$\widehat{\Psi}^{\mathrm{p}}(\alpha, \nabla_2\,\alpha, \theta) = \int_0^{\alpha} \widehat{y}(\tilde{\alpha}, \theta)\,\mathrm{d}\tilde{\alpha} + y_0(\theta)\,\frac{l_{\mathrm{p}}^2}{2}\,\|\nabla\alpha\|^2 . \tag{3.17}$$

As described in [2], l_{p} is a plastic length scale associated with a strain-gradient hardening effect that takes size effects into account in order to resolve the nonphysical mesh sensitivity of localized plastic deformation in softening materials. The temperature-dependent isotropic local hardening function $\widehat{y}(\alpha, \theta)$ can be determined experimentally for the respective material. In the context of metal plasticity we use this particular saturation-type function outlined in [112]

$$\widehat{y}(\alpha, \theta) = y_{\infty}(\theta) - (y_{\infty}(\theta) - y_0(\theta))\exp[-\omega_{\mathrm{p}}\alpha] + h(\theta)\,\alpha, \tag{3.18}$$

with the three temperature-dependent material parameters $y_0 > 0$, $y_{\infty} \geq y_0$ and $h \geq 0$ defined as

$$\begin{aligned} y_0(\theta) &= y_0(\theta_0)\,(1 - \omega_0(\theta - \theta_0)), \\ h(\theta) &= h(\theta_0)\,(1 - \omega_{\mathrm{h}}(\theta - \theta_0)), \\ y_{\infty}(\theta) &= y_{\infty}(\theta_0)\,(1 - \omega_{\mathrm{h}}(\theta - \theta_0)). \end{aligned} \tag{3.19}$$

Here, ω_{h} is the hardening/softening parameter, ω_0 is the flow stress softening parameter and ω_{p} is the saturation parameter. The initial yield stress y_0 determines the threshold of the effective elastic response. The variational derivative of $\widehat{\Psi}^{\mathrm{p}}$ with respect to α yields

$$r^{\mathrm{p}} := \delta_{\alpha}\widehat{\Psi}^{\mathrm{p}} = \partial_{\alpha}\widehat{\Psi}^{\mathrm{p}} - \mathrm{Div}[\partial_{\nabla\alpha}\widehat{\Psi}^{\mathrm{p}}] \tag{3.20}$$

reflecting the characteristics of the gradient-extended model under consideration.

Lastly, following the phase-field approach to fracture introduced in Section 2.4.3 , the phase-field fracture contribution is given in terms of the crack-density function as

$$\begin{aligned} \widehat{\Psi}^{\mathrm{f}}(\mathfrak{s}, \nabla\mathfrak{s}, \Delta\mathfrak{s}) &= \widehat{g}_{\mathrm{c}}(\alpha)\,\widehat{\gamma}(\mathfrak{s}, \nabla\mathfrak{s}, \Delta\mathfrak{s}) \\ &= \frac{\widehat{g}_{\mathrm{c}}(\alpha)}{4l_{\mathrm{f}}}\,\mathfrak{s}^2 + \frac{\widehat{g}_{\mathrm{c}}(\alpha)\,l_{\mathrm{f}}}{2}\,\nabla\mathfrak{s}\cdot\nabla\mathfrak{s} + \frac{\widehat{g}_{\mathrm{c}}(\alpha)\,l_{\mathrm{f}}^3}{4}\,(\Delta\mathfrak{s})^2. \end{aligned} \tag{3.21}$$

In the context of ductile fracture, we first require that $l_{\mathrm{p}} \geq l_{\mathrm{f}}$ such that the regularized crack zone lies inside of the plastic zone. Secondly, we propose a degradation of the Griffith-type critical energy release rate $\widehat{g}_{\mathrm{c}}$. In Miehe et al. [83], $\widehat{\Psi}^{\mathrm{p}}(\alpha, \nabla\alpha)$ is degraded as well and thus, included in the driving force of the phase-field. This behavior can be equivalently described by decreasing the local fracture energy by a factor of $1 - \frac{1}{g_c\gamma'(\mathfrak{s})} g'(\mathfrak{s})\widehat{\Psi}^{\mathrm{p}}$. Physically interpreted we can state that the critical fracture energy is decreased by the amount of energy included by the plastic work density function. Regarding the exponential character of $\widehat{\Psi}^{\mathrm{p}}$, we therefore propose the following degradation of the critical fracture energy density

$$\widehat{g}_{\mathrm{c}}(\alpha) = g_{\mathrm{c,p}} + g_{\mathrm{c,e}} \exp[-\omega_{\mathrm{f}}\alpha], \tag{3.22}$$

using the modeling parameters $\{g_{\mathrm{c,e}}, g_{\mathrm{c,p}}, \omega_{\mathrm{f}}\}$.

Finally, the crack resistance force can be obtained by the variational derivative with respect to $\mathfrak{s}$

$$r^{\mathrm{f}} := \delta_{\mathfrak{s}}\widehat{\Psi}^{\mathrm{f}} = \partial_{\mathfrak{s}}\widehat{\Psi}^{\mathrm{f}} - \mathrm{Div}[\partial_{\nabla\mathfrak{s}}\widehat{\Psi}^{\mathrm{f}}] + \Delta[\partial_{\Delta\mathfrak{s}}\widehat{\Psi}^{\mathrm{f}}]. \tag{3.23}$$

3.2.2 Dissipative response function

Regarding the porous elastoplastic material behavior, we consider a GTN type function based on [57, 123, 96], which implicitly defines the microscopic effective stress $\bar{\sigma} := \bar{\sigma}(\boldsymbol{\tau})$ in terms of the Kirchhoff stress $\boldsymbol{\tau} = \boldsymbol{\sigma}/J$ and the void volume fraction f

$$\Upsilon^{\mathrm{G}}(\boldsymbol{\tau}, \bar{\sigma}) := \frac{\sigma_{\mathrm{eq}}^2}{\bar{\sigma}^2} + 2q_1 f \cosh\left[\frac{3}{2} q_2 \frac{p}{\bar{\sigma}}\right] - \left(1 + (q_1 f)^2\right) = 0, \tag{3.24}$$

Here, $\sigma_{\mathrm{eq}} = \sqrt{3/2}\,\|\boldsymbol{\tau}_{\mathrm{dev}}/J\|$ represents the von Mises equivalent stress, whereas $p = \frac{1}{3}\mathrm{tr}[\boldsymbol{\tau}/J]$ denotes the local pressure along with the growth-based void volume fraction f and fitting parameters $q_1 \approx 1.5$ and $q_2 \approx 1.0$, see [6, 85]. Note that for $q_1 = 0$ the influence of the pressure and the void volume fraction vanishes, i.e. in this case the Gurson-type yield criterion is identical with the classical von Mises yield criterion with $\bar{\sigma} = \sigma_{\mathrm{eq}}$. Focusing on void growth and thereby neglecting other influences such as void nucleation or void softening due to shear, the evolution form of the void growth reads $\dot{f} = (1 - f)\mathrm{tr}[\boldsymbol{d}^{\mathrm{p}}]$. In line with [85], the current void volume fraction is given in terms of the plastic deformation as

$$f = 1 - \frac{1 - f_0}{J^{\mathrm{p}}} \tag{3.25}$$

with the initial void volume fraction f_0. With the effective stress $\bar{\sigma}(\boldsymbol{\tau})$ and the dissipative resistance force r^{p} we define the plastic yield function as

$$\widehat{\Phi}^{\mathrm{p}}(\bar{\sigma}(\boldsymbol{\tau}), r^{\mathrm{p}}) = \bar{\sigma}(\boldsymbol{\tau}) - r^{\mathrm{p}}. \tag{3.26}$$

A plastic Lagrange multiplier λ^{p} is introduced to enforce the Karush-Kuhn-Tucker conditions

$$\lambda^{\mathrm{p}} \geq 0, \quad \widehat{\Phi}^{\mathrm{p}} \leq 0, \quad \lambda^{\mathrm{p}}\widehat{\Phi}^{\mathrm{p}} = 0. \tag{3.27}$$

For the incorporation of the fracture mechanical behavior, we define crack threshold function as

$$\widehat{\Phi}^{\mathrm{f}}(\mathcal{H}-r^{\mathrm{f}}) = \mathcal{H} - r^{\mathrm{f}}, \tag{3.28}$$

where the energetic driving force $\mathcal{H}$ is bounded by the crack resistance force r^{f} dual to the fracture phase-field $\mathfrak{s}$. Similar to plasticity, we introduce a fracture Lagrange multiplier λ^{f} to enforce the Karush-Kuhn-Tucker conditions

$$\lambda^{\mathrm{f}} \geq 0, \quad \widehat{\Phi}^{\mathrm{f}} \leq 0, \quad \lambda^{\mathrm{f}}\,\widehat{\Phi}^{\mathrm{f}} = 0. \tag{3.29}$$

3.2.3 Evolution equations

The evolution of the thermo-elastic parts of the energetic function is given by

$$\frac{\mathrm{d}}{\mathrm{d}t}\widehat{\Psi}^{\mathrm{e},\theta} := \frac{\mathrm{d}}{\mathrm{d}t}\left[\widehat{\Psi}^{\mathrm{e}} + \widehat{\Psi}^{\theta}\right] = \frac{\partial \widehat{\Psi}^{\mathrm{e}}}{\partial \boldsymbol{b}^{\mathrm{e}}} : \dot{\boldsymbol{b}}^{\mathrm{e}} + \frac{\partial \widehat{\Psi}^{\mathrm{e}}}{\partial \mathfrak{s}}\dot{\mathfrak{s}} + \frac{\partial\left(\widehat{\Psi}^{\mathrm{e}} + \widehat{\Psi}^{\theta}\right)}{\partial \theta}\dot{\theta}, \tag{3.30}$$

with the material time derivative

$$\dot{\boldsymbol{b}}^{\mathrm{e}} = \boldsymbol{l}\boldsymbol{b}^{\mathrm{e}} + \boldsymbol{b}^{\mathrm{e}}\boldsymbol{l}^{\mathrm{T}} + \boldsymbol{F}(\dot{\boldsymbol{C}}^{\mathrm{p}})^{-1}\boldsymbol{F}^{\mathrm{T}}, \quad \dot{\boldsymbol{C}}^{\mathrm{p}} = (\dot{\boldsymbol{F}}^{\mathrm{p}})^{\mathrm{T}}\boldsymbol{F}^{\mathrm{p}} + (\boldsymbol{F}^{\mathrm{p}})^{\mathrm{T}}\dot{\boldsymbol{F}}^{\mathrm{p}}. \tag{3.31}$$

In case of isotropy, the skew-symmetric part of the spatial velocity gradient vanishes, i.e. $\boldsymbol{l} = \boldsymbol{d}$, and $\partial_{\boldsymbol{b}^{\mathrm{e}}}\widehat{\Psi}^{\mathrm{e}}$ commutes with $\boldsymbol{b}^{\mathrm{e}}$ such that the first term in (3.30) can be written as

$$\begin{aligned}\frac{\partial \widehat{\Psi}^{\mathrm{e}}}{\partial \boldsymbol{b}^{\mathrm{e}}} : \dot{\boldsymbol{b}}^{\mathrm{e}} &= \left[\frac{\partial \widehat{\Psi}^{\mathrm{e}}}{\partial \boldsymbol{b}^{\mathrm{e}}}\boldsymbol{b}^{\mathrm{e}}\right] : \boldsymbol{d} + \left[\boldsymbol{b}^{\mathrm{e}}\frac{\partial \widehat{\Psi}^{\mathrm{e}}}{\partial \boldsymbol{b}^{\mathrm{e}}}\right] : \boldsymbol{d} + \left[\frac{\partial \widehat{\Psi}^{\mathrm{e}}}{\partial \boldsymbol{b}^{\mathrm{e}}}\boldsymbol{b}^{\mathrm{e}}\right] : [\boldsymbol{F}(\dot{\boldsymbol{C}}^{\mathrm{p}})^{-1}\boldsymbol{F}^{\mathrm{T}}(\boldsymbol{b}^{\mathrm{e}})^{-1}] \\ &= \left[2\frac{\partial \widehat{\Psi}^{\mathrm{e}}}{\partial \boldsymbol{b}^{\mathrm{e}}}\boldsymbol{b}^{\mathrm{e}}\right] : [\boldsymbol{d} - \boldsymbol{d}^{\mathrm{p}}],\end{aligned} \tag{3.32}$$

where $\boldsymbol{d}^{\mathrm{p}} = -\frac{1}{2}\boldsymbol{F}(\dot{\boldsymbol{C}}^{\mathrm{p}})^{-1}\boldsymbol{F}^{\mathrm{T}}(\boldsymbol{b}^{\mathrm{e}})^{-1}$ denotes the Eulerian plastic rate of deformation tensor.

Regarding the partial derivatives in (3.30), we introduce first relations related to the Kirchhoff stress

$$\begin{aligned}\boldsymbol{\tau} &= \boldsymbol{\tau}_{\mathrm{dev}} + \boldsymbol{\tau}_{\mathrm{vol}} \\ &= 2\left(\frac{\partial \widehat{\Psi}^{\mathrm{e}}_{\mathrm{dev}}}{\partial \boldsymbol{b}^{\mathrm{e}}} + \frac{\partial \widehat{\Psi}^{\mathrm{e}}_{\mathrm{vol}}}{\partial \boldsymbol{b}^{\mathrm{e}}}\right)\boldsymbol{b}^{\mathrm{e}},\end{aligned} \tag{3.33}$$

along with the driving force of the crack phase-field

$$\mathcal{H} = -\frac{\partial \widehat{\Psi}^{\mathrm{e}}}{\partial \mathfrak{s}}, \tag{3.34}$$

and the specific entropy

$$\eta = -\frac{\partial\left(\widehat{\Psi}^{\mathrm{e}} + \widehat{\Psi}^{\theta}\right)}{\partial \theta}. \tag{3.35}$$

Based on the concept of maximum dissipation, we define an extended dissipation potential and obtain a constrained optimization problem as

$$\widehat{V} := \underbrace{\sup}_{\boldsymbol{\tau}, r^{\mathrm{p}}, \mathcal{H}-r^{\mathrm{f}}} \underbrace{\sup}_{\lambda^{\mathrm{p}}, \lambda^{\mathrm{f}}} \left[\boldsymbol{\tau} : \boldsymbol{d}^{\mathrm{p}} - J(1-f) r^{\mathrm{p}} \dot{\alpha} + (\mathcal{H} - r^{\mathrm{f}}) \dot{\mathfrak{s}} - \lambda^{\mathrm{p}} \widehat{\Phi}^{\mathrm{p}}(\boldsymbol{\tau}, r^{\mathrm{p}}) - \lambda^{\mathrm{f}} \widehat{\Phi}^{\mathrm{f}}(\mathcal{H} - r^{\mathrm{f}}) \right], \tag{3.36}$$

where the Lagrange parameters λ^{p} and λ^{f} control the non-smooth evolution of the plasticity and the fracture, respectively. This allows us to define the associated plastic evolution equations as follows

$$\boldsymbol{d}^{\mathrm{p}} = \lambda^{\mathrm{p}} \frac{\partial \widehat{\Phi}^{\mathrm{p}}}{\partial \boldsymbol{\tau}} \quad \text{and} \quad \dot{\alpha} = -\frac{\lambda^{\mathrm{p}}}{J(1-f)} \frac{\partial \widehat{\Phi}^{\mathrm{p}}}{\partial r^{\mathrm{p}}}, \tag{3.37}$$

as well as the evolution equation for the crack phase-field as

$$\dot{\mathfrak{s}} = \lambda^{\mathrm{f}} \frac{\partial \widehat{\Phi}^{\mathrm{f}}}{\partial (\mathcal{H} - r^{\mathrm{f}})} \tag{3.38}$$

along with the loading-unloading conditions introduced in (3.27) and (3.29). A penalty regularization of the Lagrange multipliers can be utilized as follows

$$\lambda^{\mathrm{p}} = \frac{1}{\eta_{\mathrm{p}}} \left\langle \widehat{\Phi}^{\mathrm{p}}(\boldsymbol{\tau}, r^{\mathrm{p}}) \right\rangle \geq 0 \quad \text{and} \quad \lambda^{\mathrm{f}} = \frac{1}{\eta_{\mathrm{f}}} \left\langle \widehat{\Phi}^{\mathrm{f}}(\mathcal{H} - r^{\mathrm{f}}) \right\rangle \geq 0, \tag{3.39}$$

where η_{p} and η_{f} are additional material parameters which characterize the viscosity of the plastic deformation and the crack propagation. Moreover, the Macaulay brackets are defined by $\langle x \rangle := (x + |x|)/2$. Eventually, we define the internal dissipation density function $\mathcal{D}_{\mathrm{int}}$ as follows

$$\mathcal{D}_{\mathrm{int}} := \nu_{\mathrm{p}} \, \boldsymbol{\tau} : \boldsymbol{d}^{\mathrm{p}} + \nu_{\mathrm{f}} \, \mathcal{H} \dot{\mathfrak{s}}, \tag{3.40}$$

where ν_{p} is a constant dissipation factor typically chosen in the range of 85% to 95% in the context of thermoplasticity, see [112, 130, 75]. In addition, ν_{f} is introduced as fracture dissipation factor based on the discussions related to an energy transfer into the thermal field in [38, 107] and the references therein.

3.2.4 Strong and weak form

The strong form of the coupled problem is summarized in Table 3.1.

1. Stress equilibrium
$$\operatorname{div}[\boldsymbol{\sigma}] + \boldsymbol{b} = \boldsymbol{0} \tag{3.41}$$
2. Energy balance
$$\theta\dot{\eta} + \operatorname{Div}[\boldsymbol{Q}] - \mathcal{D}_{\text{int}} - \mathcal{R} = 0 \tag{3.42}$$
3. Fracture force
$$\delta_{\mathfrak{s}}\widehat{\Psi} + (\mathcal{H} - r^{\text{f}}) = 0 \tag{3.43}$$
4. Hardening force
$$\delta_{\alpha}\widehat{\Psi}^{\text{p}} - r^{\text{p}} = 0 \tag{3.44}$$
5. Plastic force
$$-2\frac{\partial\widehat{\Psi}^{\text{e}}}{\partial \boldsymbol{b}^{\text{e}}}\boldsymbol{b}^{\text{e}} + \boldsymbol{\tau} = \boldsymbol{0} \tag{3.45}$$
6. Plastic strain
$$\boldsymbol{d}^{\text{p}} - \lambda^{\text{p}}\frac{\partial\widehat{\Phi}^{\text{p}}}{\partial\boldsymbol{\tau}} = \boldsymbol{0}, \quad \lambda^{\text{p}} = \frac{1}{\eta_{\text{p}}}\langle\widehat{\Phi}^{\text{p}}\rangle \tag{3.46}$$
7. Equivalent strain
$$-\dot{\alpha} - \frac{\lambda^{\text{p}}}{J(1-f)}\frac{\partial\widehat{\Phi}^{\text{p}}}{\partial r^{\text{p}}} = 0 \tag{3.47}$$
8. Phase-field
$$\dot{\mathfrak{s}} - \lambda^{\text{f}}\frac{\partial\widehat{\Phi}^{\text{f}}}{\partial(\mathcal{H} - r^{\text{f}})} = 0, \quad \lambda^{\text{f}} = \frac{1}{\eta_{\text{f}}}\langle\widehat{\Phi}^{\text{f}}\rangle \tag{3.48}$$
9. Dirichlet and Neumann conditions
$$\begin{aligned}
&\boldsymbol{\varphi} = \bar{\boldsymbol{\varphi}}(\boldsymbol{X},t) \text{ on } \partial\mathcal{B}_0^{\varphi_{\text{d}}} && \text{and} && \boldsymbol{\tau}\boldsymbol{F}^{-\text{T}}\boldsymbol{N} = \bar{\boldsymbol{T}}(\boldsymbol{X},t) \text{ on } \partial\mathcal{B}^{\varphi_{\text{n}}}\\
&\theta = \bar{\theta}(\boldsymbol{X},t) \text{ on } \partial\mathcal{B}_0^{\theta_{\text{d}}} && \text{and} && -\boldsymbol{Q}\cdot\boldsymbol{N} = Q_{\text{N}} \text{ on } \partial\mathcal{B}^{\theta_{\text{n}}}\\
&\mathfrak{s} = 1 \text{ on } \partial\mathcal{B}_0^{\mathfrak{s}_{\text{d}}} && \text{and} && \nabla\left(\frac{l_{\text{f}}^2}{2}\Delta\mathfrak{s} - \mathfrak{s}\right)\cdot\boldsymbol{N} = 0 \text{ on } \partial\mathcal{B}_0^{\mathfrak{s}_{\text{n}}}\\
&\alpha = 0 \text{ on } \partial\mathcal{B}_0^{\alpha_{\text{d}}} && \text{and} && \nabla\alpha\cdot\boldsymbol{N} = 0 \text{ on } \partial\mathcal{B}_0^{\alpha_{\text{n}}}
\end{aligned} \tag{3.49}$$
10. Initial conditions
$$\boldsymbol{\varphi}(\boldsymbol{X},0) = \boldsymbol{\varphi}_0, \quad \dot{\boldsymbol{\varphi}}(\boldsymbol{X},0) = \boldsymbol{v}_0, \quad \theta(\boldsymbol{X},0) = \theta_0, \quad \mathfrak{s}(\boldsymbol{X},0) = 0, \quad \alpha(\boldsymbol{X},0) = 0, \quad r^{\text{p}}(\boldsymbol{X},0) = 0 \tag{3.50}$$

Table 3.1: Strong formulation of the coupled problem.

Moreover, the admissible test functions related to an extended set of global primary fields $\mathfrak{U}$ is given as $\delta\mathfrak{U} := \{\delta\boldsymbol{\varphi}, \delta\mathfrak{s}, \delta\theta, \delta\alpha, \delta r^{\text{p}}\}$ i.e. the variations of the deformation, the crack phase-field, the absolute temperature and the isotropic hardening along with its dual driving force. Their spaces are defined as
$$\begin{aligned}
\mathcal{V}^{\varphi} &= \{\delta\boldsymbol{\varphi} \in H^1(\mathcal{B}_0)\,|\,\delta\boldsymbol{\varphi} = \boldsymbol{0} \text{ on } \partial\mathcal{B}_0^{\varphi_{\text{d}}}\},\\
\mathcal{V}^{\mathfrak{s}} &= \{\delta\mathfrak{s} \in H^2(\mathcal{B}_0)\,|\,\delta\mathfrak{s} = 0 \text{ on } \partial\mathcal{B}_0^{\mathfrak{s}_{\text{d}}} \wedge \nabla\delta\mathfrak{s}\cdot\boldsymbol{N} \text{ on } \partial\mathcal{B}_0\},\\
\mathcal{V}^{\theta} &= \{\delta\theta \in H^1(\mathcal{B}_0)\,|\,\delta\theta = 0 \text{ on } \partial\mathcal{B}_0^{\theta_{\text{d}}}\},\\
\mathcal{V}^{\alpha} &= \{\delta\alpha \in H^1(\mathcal{B}_0)\,|\,\delta\alpha = 0 \text{ on } \partial\mathcal{B}_0^{\alpha_{\text{d}}}\},\\
\mathcal{V}^{r^{\text{p}}} &= \{\delta r^{\text{p}} \in \mathcal{L}^2(\mathcal{B}_0)\},
\end{aligned} \tag{3.51}$$

where H^k and $k \geq 0$ denotes the Sobolev functional space of square integrable functions and derivatives. Note that the phase-field is required to be in $H^2(\mathcal{B}_0)$, which has substantial effects for the numerical solution of the coupled problem using finite elements. The weak form of the balance equations of the coupled phase-field approach to porous-ductile fracture in non-linear thermo-elasto-plastic solids reads

$$
\begin{aligned}
G_{\varphi} &:= \int_{\mathcal{B}_0} [\boldsymbol{\tau} : \nabla[\delta\boldsymbol{\varphi}] - \delta\boldsymbol{\varphi}\cdot\boldsymbol{B}] \, \mathrm{d}V - \int_{\partial\mathcal{B}_0^{\varphi n}} \delta\boldsymbol{\varphi}\cdot\bar{\boldsymbol{T}} \, \mathrm{d}A = 0, \\
G_{\mathfrak{s}} &:= \int_{\mathcal{B}_0} \left[\delta\mathfrak{s} \; \eta_{\mathrm{f}} \; \dot{\mathfrak{s}} - \delta\mathfrak{s} \; \chi_{\mathrm{f}} \left(\mathcal{H} - \frac{\widehat{g}_{\mathrm{c}}}{2 l_{\mathrm{f}}} \mathfrak{s} \right) + \chi_{\mathrm{f}} \; \widehat{g}_{\mathrm{c}} \; l_{\mathrm{f}} \; \nabla\mathfrak{s}\cdot\nabla\delta\mathfrak{s} + \frac{\chi_{\mathrm{f}}\widehat{g}_{\mathrm{c}} l_{\mathrm{f}}^3}{2} \; \Delta\mathfrak{s}\,\Delta\delta\mathfrak{s} \right] \mathrm{d}V = 0, \\
G_{\theta} &:= \int_{\mathcal{B}_0} [\delta\theta \; (\theta\dot{\eta} - \mathcal{D}_{\mathrm{int}} - \mathcal{R}) - \boldsymbol{Q}\cdot\nabla\delta\theta] \, \mathrm{d}V + \int_{\partial\mathcal{B}_0^{\theta n}} \delta\theta \; H \, \mathrm{d}A = 0, \\
G_{\alpha} &:= \int_{\mathcal{B}_0} \left[\delta\alpha \; (\widehat{y} - r^{\mathrm{p}}) + y_0(\theta) \; l_{\mathrm{p}}^2 \; \nabla\alpha\cdot\nabla\delta\alpha \right] \mathrm{d}V = 0, \\
G_{r^{\mathrm{p}}} &:= \int_{\mathcal{B}_0} \delta r^{\mathrm{p}} \left(\eta_{\mathrm{p}}\dot{\alpha} - \chi_{\mathrm{p}} \frac{\widehat{\Phi}^{\mathrm{p}}(\boldsymbol{\tau}, r^{\mathrm{p}})}{J(1-f)} \right) \mathrm{d}V = 0.
\end{aligned}
\tag{3.52}
$$

Here, $\boldsymbol{b}$ and $\boldsymbol{B}$ are given body forces per unit volume of the deformed and reference configuration, respectively, $\boldsymbol{N}$ is the outward unit normal vector and $\bar{\boldsymbol{T}}$ the surface traction at the boundary. Moreover, $\mathcal{R}$ represents the heat supply per unit volume and $H = \boldsymbol{Q}\cdot\boldsymbol{N}$ the heat supply on the thermal Neumann boundary. Appropriate Dirichlet conditions are given in Table 3.1, where $\bar{\boldsymbol{\varphi}}$ denotes a prescribed deformation and $\bar{\theta}$ a prescribed temperature. Additionally, the Dirichlet boundary condition for the crack phase-field is given by

$$
\mathfrak{s} = 1 \quad \text{on} \quad \partial\mathcal{B}_0^{\mathfrak{s}\mathrm{d}} \in \Gamma_l, \tag{3.53}
$$

ensuring that a broken state remains broken. The Karush-Kuhn-Tucker conditions in (3.27) and (3.29) are evaluated by inserting local variables defined as

$$
\chi_{\mathrm{f}} =: \begin{cases} 1 & \text{for} \quad \widehat{\Phi}^{\mathrm{f}} > 0 \\ 0 & \text{otherwise} \end{cases} \quad \text{and} \quad \chi_{\mathrm{p}} =: \begin{cases} 1 & \text{for} \quad \widehat{\Phi}^{\mathrm{p}} > 0 \\ 0 & \text{otherwise} \end{cases} . \tag{3.54}
$$

Note that by using the local variable χ_{f} in $(3.52)_2$, we demand $\dot{\mathfrak{s}} \geq 0$ for thermodynamical consistency, avoiding a transfer of energy from the fracture phase-field into the mechanical field. This prevents healing effects, which may be taken into account as well. We can also set $\chi_{\mathrm{f}} \equiv 1$ and restrict only the fully broken state, i.e. we allow for healing until the phase-field reaches one.

3.3 Spatial discretization

Following the isogeometric analysis approach presented in Section 2.5.2 and the approximations of the deformed geometry $\boldsymbol{\varphi}$ and its variation $\delta\boldsymbol{\varphi}$ defined there, the approximations of the phase-field $\mathfrak{s}$ and the temperature field θ as well as their variations $\delta\mathfrak{s}$ and $\delta\theta$ read

$$\mathfrak{s}^{\mathrm{h}} = \sum_{A\in\mathcal{I}} R^A \mathfrak{s}_A, \quad \delta\mathfrak{s}^{\mathrm{h}} = \sum_{A\in\mathcal{I}} R^A \delta\mathfrak{s}_A \tag{3.55}$$

and

$$\theta^{\mathrm{h}} = \sum_{A\in\mathcal{I}} R^A \theta_A, \quad \delta\theta^{\mathrm{h}} = \sum_{A\in\mathcal{I}} R^A \delta\theta_A. \tag{3.56}$$

Therein, we use the same spline based approximations for all fields, satisfying the required continuity $\mathfrak{s}^{\mathrm{h}}, \delta\mathfrak{s}^{\mathrm{h}} \in \mathcal{H}^2(\mathcal{B}_0)$ of the fourth-order phase-field approach, where the global shape functions $R^A : \mathcal{B}_0 \to \mathbb{R}$ are associated with control points $A \in \mathcal{I} = \{1, \ldots, \mathfrak{N}\}$ with the total number of control points $\mathfrak{N}$.

For the hardening variable α and its variation $\delta\alpha$, as well as for its dual driving force r^{p} and its variation δr^{p}, we make use of linear shape functions N^a defined on the physical mesh representation of the NURBS geometry with nodes $a \in \mathcal{J} = \{1, \ldots, \mathfrak{n}\}$ with the corresponding number of control points $\mathfrak{n}$:

$$\alpha^{\mathrm{h}} = \sum_{a\in\mathcal{J}} N^a \alpha_a, \quad \delta\alpha^{\mathrm{h}} = \sum_{a\in\mathcal{J}} N^a \delta\alpha_a \tag{3.57}$$

and

$$r^{\mathrm{p,h}} = \sum_{a\in\mathcal{J}} N^a r^{\mathrm{p}}_a, \quad \delta r^{\mathrm{p,h}} = \sum_{a\in\mathcal{J}} N^a \delta r^{\mathrm{p}}_a. \tag{3.58}$$

The more natural choice of approximating the two plastic fields with the same quadratic NURBS shape functions leads to oscillations within both fields, indicating stability issues as already observed in [7]. The above described scheme has shown to be stable and numerically robust within our numerical examples.

Inserting the approximations into (3.52) yields the semi-discrete set of coupled equations

$$
\begin{aligned}
G^{\mathrm{h}}_{\varphi} &:= \delta \boldsymbol{q}_A \cdot \left[\int_{\mathcal{B}_0} \boldsymbol{\tau}^{\mathrm{h}} \nabla [R^A] \, \mathrm{d}V - \boldsymbol{F}^{\mathrm{ext},A} \right], \\
G^{\mathrm{h}}_{\mathfrak{s}} &:= \delta \mathfrak{s}_A \left[M^{AB}_{\mathfrak{s}} \dot{\mathfrak{s}}_B - \int_{\mathcal{B}_0} R^A \mathcal{H}^{\mathrm{h}} \, \mathrm{d}V + K^{AB}_{\mathfrak{s}} \mathfrak{s}_B \right], \\
G^{\mathrm{h}}_{\theta} &:= \delta \theta_A \left[\int_{\mathcal{B}_0} \left(\dot{\eta}^{\mathrm{h}} R^A R^B \theta_B - R^A \mathcal{D}^{\mathrm{h}}_{\mathrm{int}} - \nabla R^A \boldsymbol{Q}^{\mathrm{h}} \right) \mathrm{d}V - Q^{\mathrm{ext},A} \right], \\
G^{\mathrm{h}}_{\alpha} &:= \delta \alpha_a \left[\int_{\mathcal{B}_0} N^a (\widehat{y}^{\mathrm{h}} - N^b r^{\mathrm{p}}_b) \, \mathrm{d}V + K^{ab}_{\alpha} \alpha_b \right], \\
G^{\mathrm{h}}_{r^{\mathrm{p}}} &:= \delta r^{\mathrm{p}}_a \left[M^{ab}_{r^{\mathrm{p}}} \dot{\alpha}_b - \int_{\mathcal{B}_0} \chi_{\mathrm{p}} N^a \, \frac{\widehat{\Phi}^{\mathrm{p,h}}(\boldsymbol{\tau}^{\mathrm{h}}, r^{\mathrm{p,h}})}{J^{\mathrm{h}}(1 - f^{\mathrm{h}})} \, \mathrm{d}V \right].
\end{aligned}
\tag{3.59}
$$

The Kirchhoff stress tensor, the local entropy and the phase-field driving force are given by

$$
\boldsymbol{\tau}^{\mathrm{h}} = 2 \frac{\partial \widehat{\Psi}^{\mathrm{e,h}}(\bar{\bar{\boldsymbol{b}}}^{\mathrm{e,h}}, \mathfrak{s}^{\mathrm{h}}, \theta^{\mathrm{h}})}{\partial \boldsymbol{b}^{\mathrm{e,h}}} \boldsymbol{b}^{\mathrm{e,h}}, \quad \mathcal{H}^{\mathrm{h}} = - \frac{\partial \widehat{\Psi}^{\mathrm{e,h}}(\bar{\bar{\boldsymbol{b}}}^{\mathrm{e,h}}, \mathfrak{s}^{\mathrm{h}}, \theta^{\mathrm{h}})}{\partial \mathfrak{s}^{\mathrm{h}}} \quad \text{and} \quad \eta^{\mathrm{h}} = - \frac{\partial \widehat{\Psi}^{\mathrm{e},\theta,\mathrm{h}}(\bar{\bar{\boldsymbol{b}}}^{\mathrm{e,h}}, \mathfrak{s}^{\mathrm{h}}, \theta^{\mathrm{h}})}{\partial \theta^{\mathrm{h}}},
\tag{3.60}
$$

respectively. The semi-discrete definition of the heat flux vector along with the conductivity tensor read

$$
\boldsymbol{Q}^{\mathrm{h}} = \boldsymbol{K}^{\mathrm{h}} \nabla R^A \theta_A \quad \text{and} \quad \boldsymbol{K}^{\mathrm{h}} = [K_0 (1 - \omega_{\mathrm{K}} (\theta^{\mathrm{h}} - \theta_0)) (1 - \mathfrak{s}^{\mathrm{h}}) + K^{\mathrm{conv}} \mathfrak{s}^{\mathrm{h}}] (\boldsymbol{C}^{\mathrm{h}})^{-1}, \tag{3.61}
$$

where

$$
\boldsymbol{C}^{\mathrm{h}} = \boldsymbol{q}_A \cdot \boldsymbol{q}_B \nabla R^A \otimes \nabla R^B. \tag{3.62}
$$

Moreover, the semi-discrete external contributions in $(3.59)_1$ and $(3.59)_3$ are formulated as

$$
\boldsymbol{F}^{\mathrm{ext},A} = \int_{\mathcal{B}_0} R^A \boldsymbol{B} \, \mathrm{d}V + \int_{\partial \mathcal{B}_0^{\varphi_{\mathrm{n}}}} R^A \bar{\boldsymbol{T}} \, \mathrm{d}A \quad \text{and} \quad Q^{\mathrm{ext},A} = \int_{\mathcal{B}_0} R^A \mathcal{R} \, \mathrm{d}V + \int_{\partial \mathcal{B}_0^{\theta_{\mathrm{n}}}} R^A Q_{\mathrm{N}} \, \mathrm{d}A.
\tag{3.63}
$$

The coefficients of the matrices in $(3.59)_2$ are given by

$$
\begin{aligned}
M^{AB}_{\mathfrak{s}} &= \eta_{\mathrm{f}} \int_{\mathcal{B}_0} R^A R^B \, \mathrm{d}V, \\
K^{AB}_{\mathfrak{s}} &= \frac{1}{l_{\mathrm{f}}} \int_{\mathcal{B}_0} g^{\mathrm{h}}_{\mathrm{c}} \chi_{\mathrm{f}} \left(\frac{1}{2} R^A R^B + l_{\mathrm{f}}^2 \nabla R^A \cdot \nabla R^B + \frac{l_{\mathrm{f}}^4}{2} \Delta R^A \Delta R^B \right) \mathrm{d}V,
\end{aligned}
\tag{3.64}
$$

whereas the matrices in $(3.59)_4$ and $(3.59)_5$ take the form

$$K_{\alpha}^{ab} = y_0 l_{\mathrm{p}}^2 \int_{\mathcal{B}_0} \nabla N^a \cdot \nabla N^b \,\mathrm{d}V \quad \text{and} \quad M_{r^{\mathrm{p}}}^{ab} = \eta_{\mathrm{p}} \int_{\mathcal{B}_0} N^a N^b \,\mathrm{d}V. \tag{3.65}$$

The semi-discrete definitions of the critical fracture energy density and the local hardening function read

$$g_{\mathrm{c}}^{\mathrm{h}} = g_{\mathrm{c,p}} + g_{\mathrm{c,e}} \exp[-\omega_{\mathrm{f}} \alpha^{\mathrm{h}}] \tag{3.66}$$

and

$$\widehat{y}^{\mathrm{h}} = y_{\infty}(\theta^{\mathrm{h}}) - (y_{\infty}(\theta^{\mathrm{h}}) - y_0(\theta^{\mathrm{h}})) \exp[-\omega_{\mathrm{p}} \alpha^{\mathrm{h}}] + h(\theta^{\mathrm{h}})\, \alpha^{\mathrm{h}}, \tag{3.67}$$

respectively. The discrete void volume fraction optained from $f^{\mathrm{h}} = 1 - (1 - f_0)/J^{\mathrm{p,h}}$ with $J^{\mathrm{p,h}} = \sqrt{\det[\boldsymbol{C}^{\mathrm{p,h}}]}$ and J^{h} the discrete version of the Jacobian determinant. Eventually, the internal dissipation in semi-discrete form reads

$$\mathcal{D}_{\mathrm{int}}^{\mathrm{h}} := \nu_{\mathrm{p}}\, \boldsymbol{\tau}^{\mathrm{h}} : \boldsymbol{d}^{\mathrm{p,h}} + \nu_{\mathrm{f}}\, \mathcal{H}^{\mathrm{h}} \dot{\mathfrak{s}}^{\mathrm{h}}, \tag{3.68}$$

using the semi-discrete evaluation of the plastic deformation rate tensor $\boldsymbol{d}^{\mathrm{p,h}}$.

3.4 Temporal discretization

Finally, in order to obtain a set of non-linear algebraic equations to be solved via a Newton-Raphson method, we discretize the semi-discrete coupled problem (3.59) in time. Therefore, we subdivide the considered time interval $\mathcal{T}$ into a sequence of times $t_0, \ldots, t_n, t_{n+1}, \ldots T$, where $(\bullet)_n$ and $(\bullet)_{n+1}$ denote the value of a given physical quantity at time t_n and t_{n+1}, respectively. Suppose that the discrete set of state variables at t_n given by $\mathfrak{U}_{A,n} := \{\boldsymbol{q}_{A,n}, \mathfrak{s}_{A,n}, \theta_{A,n}, \alpha_{A,n}, r_{A,n}^{\mathrm{p}}\}$ and the local plastic deformation variable $\boldsymbol{C}_n^{\mathrm{p,h}}$ at time t_n are known and the time step size $\Delta t = t_{n+1} - t_n$ is given. Then, the goal is to determine the corresponding fields at time t_{n+1} via the algorithmic approximation to the weak formulation (3.59) defined as

$$
\begin{aligned}
G_{\varphi}^{\mathrm{h},n+1} &:= \delta \boldsymbol{q}_A \cdot \left[\int_{\mathcal{B}_0} \boldsymbol{\tau}_{n+1}^{\mathrm{h}} \nabla[R^A] \, \mathrm{d}V - \boldsymbol{F}_{n+1}^{\mathrm{ext},A} \right], \\
G_{\mathfrak{s}}^{\mathrm{h},n+1} &:= \delta \mathfrak{s}_A \left[M_{\mathfrak{s}}^{AB} \frac{\mathfrak{s}_{B,n+1} - \mathfrak{s}_{B,n}}{\Delta t} - \int_{\mathcal{B}_0} R^A \mathcal{H}_{n+1}^{\mathrm{h}} \, \mathrm{d}V + K_{\mathfrak{s},n+1}^{AB} \mathfrak{s}_{B,n+1} \right], \\
G_{\theta}^{\mathrm{h},n+1} &:= \delta \theta_A \left[\int_{\mathcal{B}_0} \left(\frac{\eta_{n+1}^{\mathrm{h}} - \eta_n^{\mathrm{h}}}{\Delta t} R^A R^B \theta_{B,n+1} - R^A \mathcal{D}_{\mathrm{int},n+1}^{\mathrm{h}} - \nabla R^A \boldsymbol{Q}_{n+1}^{\mathrm{h}} \right) \mathrm{d}V - Q_{n+1}^{\mathrm{ext,A}} \right], \\
G_{\alpha}^{\mathrm{h},n+1} &:= \delta \alpha_a \left[\int_{\mathcal{B}_0} N^a (\widehat{y}_{n+1}^{\mathrm{h}} - N^b r_{b,n+1}^{\mathrm{p}}) \, \mathrm{d}V + K_{\alpha}^{ab} \alpha_{b,n+1} \right], \\
G_{r^{\mathrm{p}}}^{\mathrm{h},n+1} &:= \delta r_a^{\mathrm{p}} \left[M_{r^{\mathrm{p}}}^{ab} \frac{\alpha_{b,n+1} - \alpha_{b,n}}{\Delta t} - \int_{\mathcal{B}_0} \chi_{\mathrm{p},n+1} N^a \frac{\widehat{\Phi}_{n+1}^{\mathrm{p,h}}(\boldsymbol{\tau}_{n+1}^{\mathrm{h}}, r_{n+1}^{\mathrm{p,h}})}{J_{n+1}^{\mathrm{h}}(1 - f_{n+1}^{\mathrm{h}})} \, \mathrm{d}V \right].
\end{aligned}
\tag{3.69}
$$

Hereby, a full-discrete definition of the internal dissipation is given by

$$
\mathcal{D}_{\mathrm{int},n+1}^{\mathrm{h}} := \nu_{\mathrm{p}} \, \boldsymbol{\tau}_{n+1}^{\mathrm{h}} : \boldsymbol{d}_{n+1}^{\mathrm{p,h}} + \nu_{\mathrm{f}} \, \mathcal{H}_{n+1}^{\mathrm{h}} \frac{\mathfrak{s}_{n+1}^{\mathrm{h}} - \mathfrak{s}_n^{\mathrm{h}}}{\Delta t}. \tag{3.70}
$$

Taking into account small values for the plastic viscosity parameter η_{p}, we propose the following approximation

$$
\boldsymbol{\tau}_{n+1}^{\mathrm{h}} : \boldsymbol{d}_{n+1}^{\mathrm{p,h}} \approx J_{n+1}^{\mathrm{h}}(1 - f_{n+1}^{\mathrm{h}}) \, r_{n+1}^{\mathrm{p,h}} \frac{\alpha_{n+1}^{\mathrm{h}} - \alpha_n^{\mathrm{h}}}{\Delta t} \tag{3.71}
$$

such that the internal dissipation can be recast as

$$
\mathcal{D}_{\mathrm{int},n+1}^{\mathrm{h}} := \nu_{\mathrm{p}} \, J_{n+1}^{\mathrm{h}}(1 - f_{n+1}^{\mathrm{h}}) \, r_{n+1}^{\mathrm{p,h}} N^a \frac{\alpha_{a,n+1} - \alpha_{a,n}}{\Delta t} + \nu_{\mathrm{f}} \, \mathcal{H}_{n+1}^{\mathrm{h}} R^A \frac{\mathfrak{s}_{A,n+1} - \mathfrak{s}_{A,n}}{\Delta t} \tag{3.72}
$$

for practical reasons.

Note that we apply a staggered scheme for the solution of the multi-field problem, i.e. the displacement field along with the plastic and hardening fields $\{\boldsymbol{q}_{A,n+1}, \boldsymbol{C}_{n+1}^{\mathrm{p,h}}, \alpha_{A,n+1}$ and $r_{A,n+1}^{\mathrm{p}}\}$, the crack phase-field $\mathfrak{s}_{A,n+1}$ and the temperature field $\theta_{A,n+1}$ are solved successively. For the time integration of the plastic evolution equations, the construction of a return mapping algorithm is most crucial. Therefore, we define a trial state as *

$$
\boldsymbol{b}_{\mathrm{tr}}^{\mathrm{e}} = \boldsymbol{F}_{n+1}(\boldsymbol{C}^{\mathrm{p}})_n^{-1} \boldsymbol{F}_{n+1}^{\mathrm{T}} \tag{3.73}
$$

assuming that no further plastic deformation occurs within the time step. Based on this trial state, we evaluate the yield criteria (6.37). If $\widehat{\Phi}_{\mathrm{tr}}^{\mathrm{p}} \leq 0$, then the process is purely

* For the sake of readability, we neglect the labeling of the spatial approximation in the following.

elastic and the elastic trial state is the solution. If on the other hand $\widehat{\Phi}^{\mathrm{p}}_{\mathrm{tr}} > 0$, then the trial state is not admissible and a plastic correction is required. Therefore, we apply an exponential integration scheme regarding (3.37) which leads to

$$(\boldsymbol{C}^{\mathrm{p}})^{-1}_{n+1} = \exp\big[-2\Delta t\lambda^{\mathrm{p}}_{n+1}\boldsymbol{F}^{-1}_{n+1}\boldsymbol{n}_{n+1}\boldsymbol{F}_{n+1}\big](\boldsymbol{C}^{\mathrm{p}})^{-1}_{n}, \quad \boldsymbol{n}_{n+1} = \frac{\partial\widehat{\Phi}^{\mathrm{p}}_{n+1}}{\partial\boldsymbol{\tau}_{n+1}} \tag{3.74}$$

and

$$(\lambda^{\mathrm{e}}_{a,n+1})^2 = (\lambda^{\mathrm{e}}_{a,\mathrm{tr}})^2\exp\big[-2\Delta t\lambda^{\mathrm{p}}_{n+1}n_{a,n+1}\big], \tag{3.75}$$

respectively. Note that in contrast to standard von Mises plasticity $\boldsymbol{n}_{n+1} \neq \boldsymbol{n}_{\mathrm{tr}}$ and $\|\boldsymbol{n}_{n+1}\| \neq 1$, i.e. the plastic correction has to be performed by the Lagrange multiplier $\lambda^{\mathrm{p}}_{n+1}$ as well as the components $n_{a,n+1}$ which can be obtained by solving the non-linear relations

$$\widehat{\Phi}^{\mathrm{p}}_{n+1} - \eta_{\mathrm{p}}\lambda^{\mathrm{p}}_{n+1} = 0 \quad \text{and} \quad \frac{\partial\widehat{\Phi}^{\mathrm{p}}_{n+1}}{\partial\tau_{a,n+1}} - n_{a,n+1} = 0 \tag{3.76}$$

via an internal Newton-Raphson iteration. In addition, the void volume fraction f_{n+1} is locally calculated by

$$f_{n+1} = \max\big\{f_0, 1 - (1 - f_0)/\sqrt{\det[\boldsymbol{C}^{\mathrm{p}}_{n+1}]}\big\}. \tag{3.77}$$

An overview of the return map algorithm is given in Box 1.

$\boldsymbol{F}_{n+1}, \boldsymbol{C}^{\mathrm{p}}_n, \alpha_{n+1}, r^{\mathrm{p}}_{n+1}, \mathfrak{s}_n(\text{frozen}), \theta_n(\text{frozen})$

Compute trial state:

$\boldsymbol{b}^{\mathrm{e}}_{\mathrm{tr}} = \boldsymbol{F}_{n+1}(\boldsymbol{C}^{\mathrm{p}}_n)^{-1}\boldsymbol{F}^{\mathrm{T}}_{n+1}$

$[(\lambda^{\mathrm{e}}_{a,\mathrm{tr}})^2, \boldsymbol{n}_a] = \mathrm{eig}(\boldsymbol{b}^{\mathrm{e}}_{\mathrm{tr}})$

$f_n = \max\{f_0, 1-(1-f_0)/\sqrt{\det[\boldsymbol{C}^{\mathrm{p}}_n]}\}.$

$J^{\mathrm{e}}_{\mathrm{tr}} = \prod_a \lambda^{\mathrm{e}}_{a,\mathrm{tr}}$

$\boldsymbol{\tau}_{\mathrm{dev,tr}} = \mu g (J^{\mathrm{e}}_{\mathrm{tr}})^{-\frac{2}{3}g} \sum_a \left((\lambda^{\mathrm{e}}_{a,\mathrm{tr}})^{2g} - \frac{1}{3}\sum_i (\lambda^{\mathrm{e}}_{i,\mathrm{tr}})^{2g} \right) \boldsymbol{n}_a \otimes \boldsymbol{n}_a$

if $J^{\mathrm{e}}_{\mathrm{tr}} > 1$ **then**

| $p_{\mathrm{tr}} = \frac{\kappa}{2}(J_{n+1})^{-1}\left((J^{\mathrm{e}}_{\mathrm{tr}})^g - (J^{\mathrm{e}}_{\mathrm{tr}})^{-g} - 3\beta(\theta_n-\theta_0)(1+(J^{\mathrm{e}}_{\mathrm{tr}})^{-2g})\right) g (J^{\mathrm{e}}_{\mathrm{tr}})^g$

else

| $p_{\mathrm{tr}} = \frac{\kappa}{2}(J^{\mathrm{p}}_n)^{-1}\left(J^{\mathrm{e}}_{\mathrm{tr}} - (J^{\mathrm{e}}_{\mathrm{tr}})^{-1} - 3\beta(\theta_n-\theta_0)(1+(J^{\mathrm{e}}_{\mathrm{tr}})^{-2})\right)$

end

solve $\bar{\sigma}_{tr}$

| $\frac{3\|\boldsymbol{\tau}_{\mathrm{dev,tr}}\|^2}{2(\bar{\sigma}_{tr})^2(J_{n+1})^2} + 2q_1 f_n \cosh\left[\frac{3}{2}q_2\frac{p_{\mathrm{tr}}}{\bar{\sigma}_{tr}}\right] - \left(1+(q_1 f_n)^2\right) = 0$

$\widehat{\Phi}^{\mathrm{p}}_{\mathrm{tr}} = \bar{\sigma}_{tr} - r^{\mathrm{p}}_{n+1}$

if $\widehat{\Phi}^{\mathrm{p}}_{\mathrm{tr}} > 0$ **then**

| **solve** $\lambda^{\mathrm{p}}_{n+1}, n_{a,n+1}$

| | $\frac{3\|\boldsymbol{\tau}_{\mathrm{dev,n+1}}\|^2}{2(\bar{\sigma}_{n+1})^2(J_{n+1})^2} + 2q_1 f_n \cosh\left[\frac{3}{2}q_2\frac{p_{n+1}}{\bar{\sigma}_{n+1}}\right] - \left(1+(q_1 f_n)^2\right) = 0$

| | $\bar{\sigma}_{n+1} - r^{\mathrm{p}}_{n+1} - \eta_{\mathrm{p}}\lambda^{\mathrm{p}}_{n+1} = 0$

| | $\frac{\partial \widehat{\Phi}^{\mathrm{p}}_{n+1}}{\partial \tau_{a,n+1}} - n_{a,n+1} = 0$

| | with $(\lambda^{\mathrm{e}}_{a,n+1})^2 = (\lambda^{\mathrm{e}}_{a,\mathrm{tr}})^2 \exp\left[-2\Delta t \lambda^{\mathrm{p}}_{n+1} n_{a,n+1}\right]$

| Compute: $\lambda^{\mathrm{e}}_{a,n+1}, \boldsymbol{\tau}_{\mathrm{dev},n+1}(\lambda^{\mathrm{e}}_{a,n+1}), p_{n+1}(\lambda^{\mathrm{e}}_{a,n+1})$

| $\boldsymbol{\tau}_{n+1} = \boldsymbol{\tau}_{\mathrm{dev,n+1}} + p_{n+1}J_{n+1}\boldsymbol{I}$

else

| $\lambda^{\mathrm{p}}_{n+1} = 0$

| $\lambda^{\mathrm{e}}_{a,n+1} = \lambda^{\mathrm{e}}_{a,\mathrm{tr}}$

| $\boldsymbol{\tau}_{n+1} = \boldsymbol{\tau}_{\mathrm{dev,tr}} + p_{\mathrm{tr}}J_{n+1}\boldsymbol{I}$

end

Box 1: Return mapping algorithm.

3.5 Numerical examples

In this section we demonstrate the accuracy and performance of the newly developed phase-field formulation for thermo-porous ductile fracture. In particular, two examples are considered. Firstly, a benchmark example with experimental data allows us to explore the full range of porous thermoplastic material behavior, so that we end up with a comprehensive validated material setting summarized in 3.2. Finally, a second example from the third Sandia Fracture Challenge is used to demonstrate the ability of the model to be applied to more complex geometries that produce three-dimensional fracture patterns.

3.5.1 Tensile test

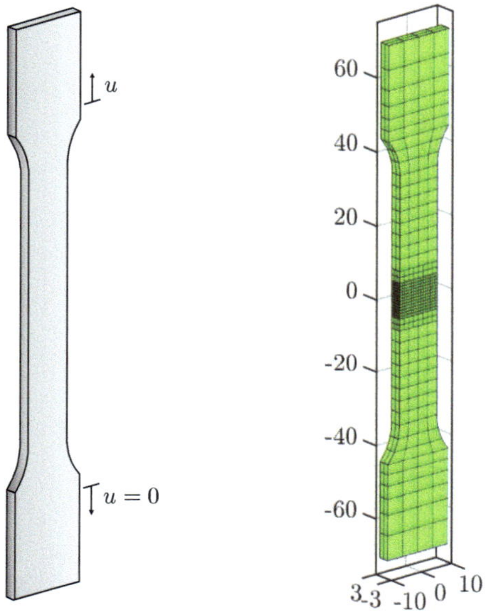

Figure 3.1: **Tensile test.** Reference configuration and computational mesh.

The following benchmark example has been presented in Ambati et al. [7], where experimental data of steel 1.0553 are used to adjust the ductile fracture model proposed therein. For the numerical investigations in this section, we address the following issues:

I For the present model, physically more comprehensive model, the results have to be in a good agreement with the experimental results in terms of hardening, necking, crack initialization, propagation and failure as well.

II For a physical reasonable formulation of ductile fracture, a non-negative dissipation of energy must be taken into account. This has to investigated for the elastoplastic behavior as well as for the fracture mechanical behavior, cf. (3.40).

III The modeling of the critical fracture energy density plays an important role for the ductile fracture behavior. Regarding the definition in (3.22), we have to investigate the effect of $g_{c,p}$ on crack initialization.

Elastic parameters	
Shear modulus μ	$73255\,\mathrm{MPa}$
Bulk modulus κ	$150000\,\mathrm{MPa}$
Plastic parameters	
Yield stress $y_0(\theta_0)$	$345.6\,\mathrm{MPa}$
Ultimate yield stress $y_\infty(\theta_0)$	$688.25\,\mathrm{MPa}$
Hardening modulus $h(\theta_0)$	$300\,\mathrm{MPa}$
Saturation exponent ω_p	16.93
Thermal softening parameter ω_h	$0.002\,\mathrm{K}^{-1}$
Thermal softening parameter ω_0	$0.002\,\mathrm{K}^{-1}$
Plastic viscosity η_p	$1\cdot 10^{-7}\,\mathrm{MPa}\cdot\mathrm{s}$
Plastic length scale l_p	$0.78125\,\mathrm{mm}$
Initial void fraction f_0	0.005
Gurson fitting parameter q_1	1.5
Gurson fitting parameter q_2	1
Phase-field fracture parameters	
Brittle critical fracture energy $g_{\mathrm{c_{e0}}}$	$825\,\mathrm{kJ/m^2}$
Ductile critical fracture energy $g_{\mathrm{c_{p0}}}$	$175\,\mathrm{kJ/m^2}$
Saturation exponent ω_f	42.325
Fracture viscosity η_f	$1\cdot 10^{-7}\,\mathrm{MPa}\cdot\mathrm{s}$
Fracture length scale l_f	$0.78125\,\mathrm{mm}$
Thermal parameters	
Specific heat capacity c	$3588\,\mathrm{kJ/(m^3\cdot K)}$
Thermal expansion coefficient β	$1\cdot 10^{-5}\,\mathrm{K}^{-1}$
Conductivity K_0	$45\,\mathrm{W/(m\cdot K)}$
Convection K_conv	$0\,\mathrm{W/(m\cdot K)}$
Conductivity softening parameter ω_K	$0\,\mathrm{K}^{-1}$
Reference temperature θ_0	$293\,\mathrm{K}$
Plastic dissipation factor ν_p	0.9
Fracture dissipation factor ν_f	0.9

Table 3.2: Material setting of the steel 1.0553 used for the numerical simulations in this section.

IV It is well known for phase-field formulations to brittle fracture, that the fracture length scale l_f has the character of a material parameter and influences the crack initialization significantly. A study of the fracture length scale parameter is indispensable, since we suppose that the effect is more pronounced for ductile fracture.

V To overcome mesh sensitivities we have proposed a gradient extended plasticity model as given in (3.17). A comparison of different plastic length scale parameters up to a local plasticity model has to be part of a comprehensive investigation of the model.

VI In [7], possible oscillations of the stress field are addressed. With regard to the four-field formulation along with the proposed approximation scheme, we have to investigate whether oscillations occur within the results.

VII For the present model, a Gurson-type yield criterion is applied taking into account the growth of micro-voids. In comparison to a standard von Mises type yield criterion, the behavior in terms of hardening, necking and crack initialization has to be investigated.

VIII A temperature-dependent formulation of the hardening behavior is incorporated within the present model which determines the elastic response such that we expect an impact on crack initialization as well. Therefore, investigations based on isothermal simulations at different temperature levels are useful to understand the model in more detail.

IX For a physically correct representation of the thermomechanical behavior, an energy transfer into the thermal field is taken into account via the internal dissipation function defined in (3.40). By conducting fully coupled simulations with different values of the dissipation factors, the interaction of the different fields can be investigated for the entire thermomechanical process.

X Due to the thermomechanical coupling, the formulation is strongly rate-dependent. To investigate this effect, we conduct further simulations and apply different deformation rates.

The benchmark problem considers a flat I-shaped specimen of size 140 mm × 20 mm × 3 mm with reduced width of 12.5 mm in the middle. The length of the flaps is 25 mm and the radius in the transition is 15 mm. Figure 3.1 shows the geometry in the reference configuration along with the applied boundary conditions and the computational mesh. The outer 20 mm of both flaps are subject to Dirichlet boundary conditions. To be specific, the lower flap is fixed in space and the upper flap is moved upwards by a displacement rate of 0.5 mm/min within a quasi-static simulation setting neglecting inertia effects. For the thermal field, no heat in- or outflow is allowed, i.e. the system is adiabatically isolated. The used computational mesh consists of in total 1936 quadratic NURBS elements including a two level local refinement of the region in which plastification, necking and crack propagation is expected.

Regarding **point I**, Figure 3.2 demonstrates a good agreement of the load deflection result obtained by an isothermal simulation at $\theta = 293\,\mathrm{K}$ and the experimental result given in [7]. Therein and in what follows, the displacement is measured at ±25 mm from the center of the specimen. Note that we obtain no significant softening effect due to the crack phase-field before crack initialization. In our opinion, this is the correct representation of

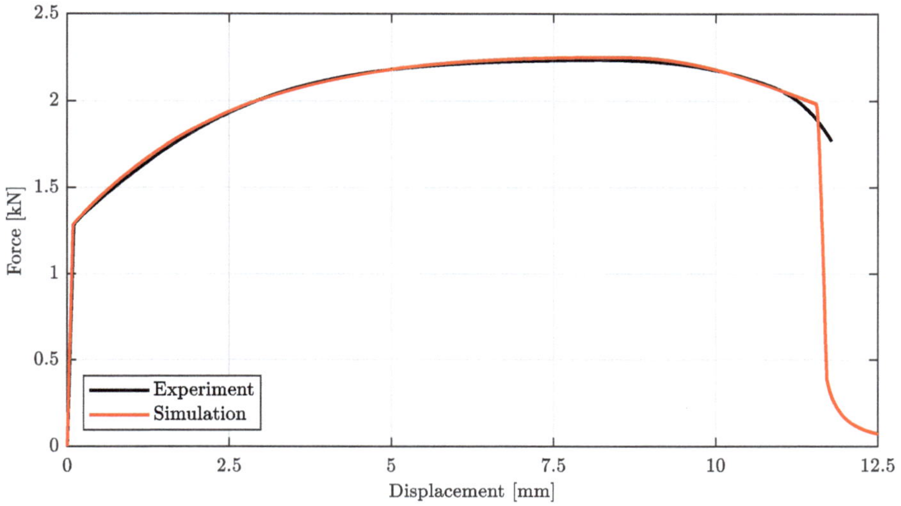

Figure 3.2: **Tensile test.** Load deflection result of the present model and the experiment given in [7].

the problem, since inertia effects are neglected within the simulations. Nevertheless, we can reproduce a softer behavior at crack initialization by increasing the fracture viscosity parameter η_f.

As shown in Figure 3.3, and in accordance with the experiment, the crack phase-field begins to evolve from the center of the specimen, driven by plastic strain localization.

Next, the non-negative energy dissipation is checked, as addressed in **item II**. In Figure 3.4 we see that both the dissipation due to plastification and the crack growth take positive values throughout the process, which is consistent with the second law of thermodynamics. As for the plastic part, once the yield stress is reached, the system dissipates energy, which is enhanced by the strain hardening behavior. In contrast, dissipation into the fracture phase-field is much lower until crack propagation. Thereafter, a sudden increase in fracture mechanical dissipation occurs up to a value of 216.5038 W and plastic dissipation ceases as dissipation of elastic energy by crack propagation leads to a lower stress level.

Concerning **point III**, we next consider the effect of different values of the reduced critical crack energy density $g_{c,p}$. In Figure 3.5, we can observe that crack initiation depends strongly on $g_{c,p}$. A change of one percent with respect to the initial critical crack energy density $g_{c,e}$ significantly shifts the point at which the material begins to crack, but has no effect on the material behavior before that. This is in agreement with the observations in Figure 3.2 and our assumption of abrupt failure of steel-like material in a quasi-static simulation environment. A similar effect can be observed by varying the fracture length scale parameter, as shown in Figure 3.6. Note that this confirms our conjecture in **item**

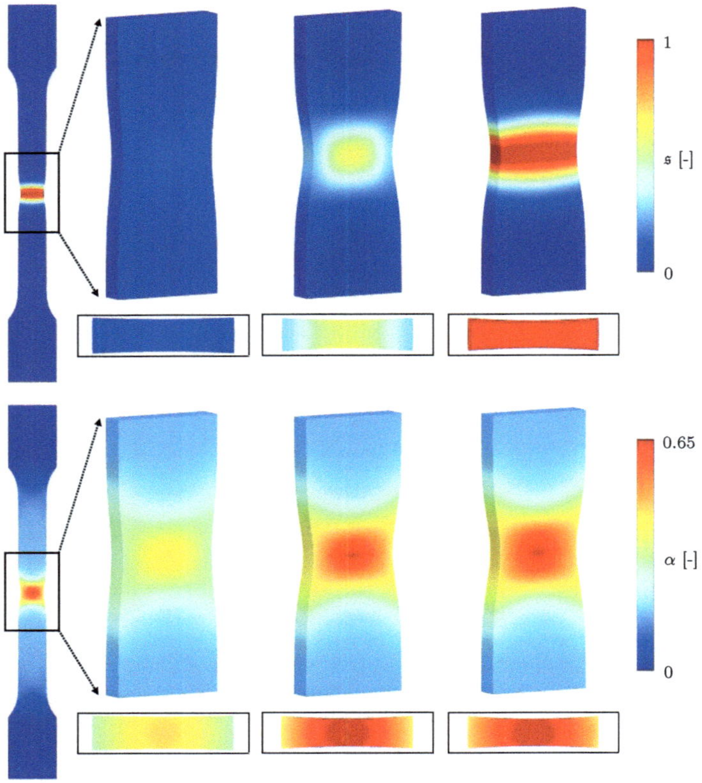

Figure 3.3: **Tensile test.** Result of an isothermal simulation at $\theta = 293\,\mathrm{K}$. The crack phase-field (first row) and the equivalent plastic strain (second row) are plotted at displacements of $u_{\pm 25} = [10.50,\, 11.6,\, 12.5]\,\mathrm{mm}$. Additionally, the constricted cross-section in the middle of specimen is shown, where the solid line represents the undeformed configuration.

IV that the length scale has a large effect on crack initiation and therefore we need to fix the value of l_f for all subsequent simulations.

Figure 3.7 shows results for different values of the plastic length scale l_p, as addressed in **point V**. As expected, the results deviate slightly when necking occurs, but since in this case the differences in plastic deformation are too small, there is no visible change with respect to crack initiation. Further numerical investigations regarding the plastic length scale in [37, 2], have shown the size effects to overcome the unphysical mesh sensitivity of the localized plastic deformation and consequently the crack initiation.

Regarding **point VII**, an investigation of the modified Gurson model with different values of the fitting parameters up to a purely isochoric model with $q_1 = q_2 = 0$ is shown in Figure 3.8 and 3.10. The von Mises stress distribution and the void volume fraction

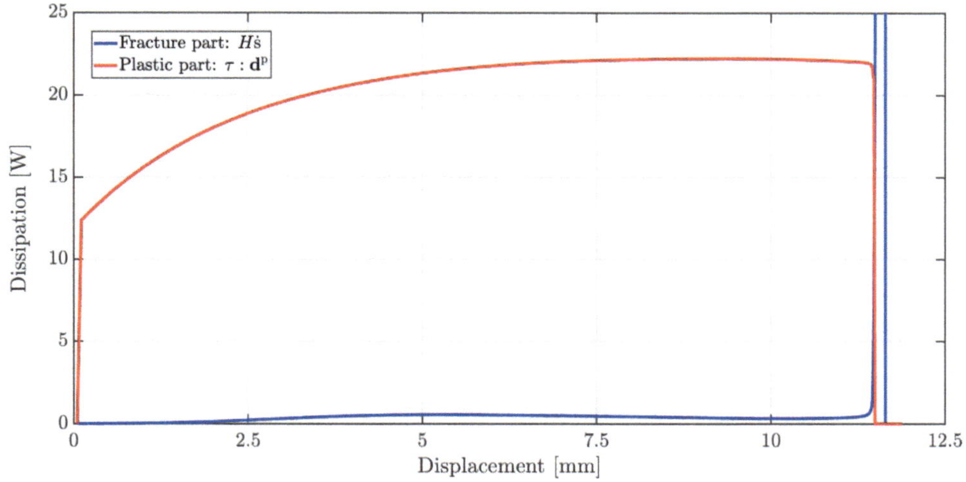

Figure 3.4: **Tensile test.** Dissipation of energy due to plastification and crack growth. The fracture mechanical dissipation part reaches a value of 216.5038 W which is not captured in the diagram for reasons of clarity.

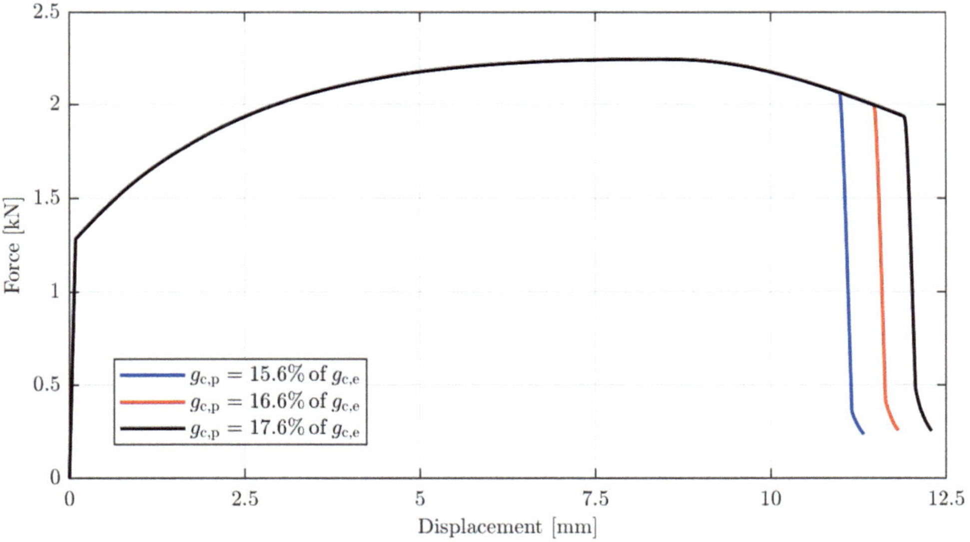

Figure 3.5: **Tensile test.** Load deflection results for different values of $g_{\mathrm{c,p}}$.

at crack initiation are shown in Figure 3.10. Here it can be observed that the necking behavior is more pronounced by increasing the Gurson parameters, i.e., by increasing the volumetric fraction within the model, which is clearly visible by the deformed cross

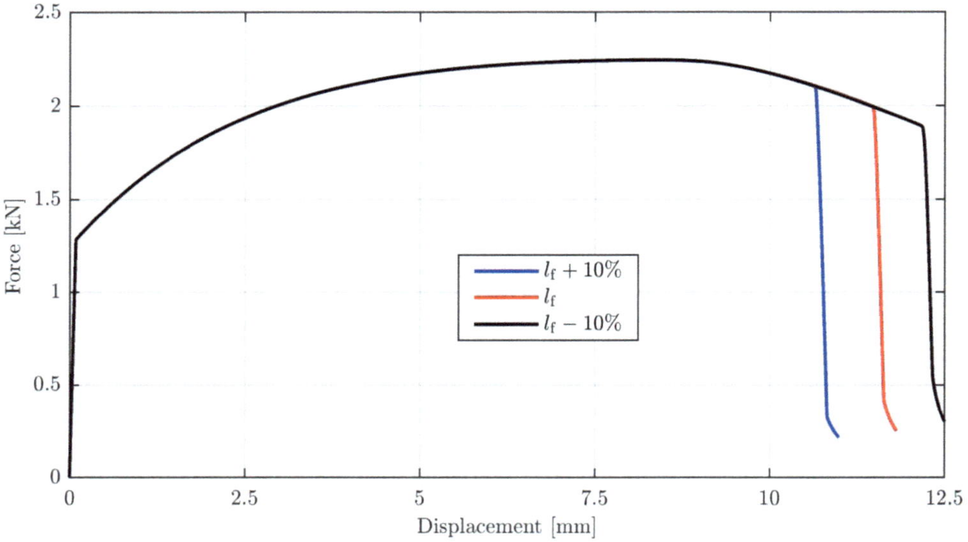

Figure 3.6: **Tensile test.** Load deflection results for different values of l_f .

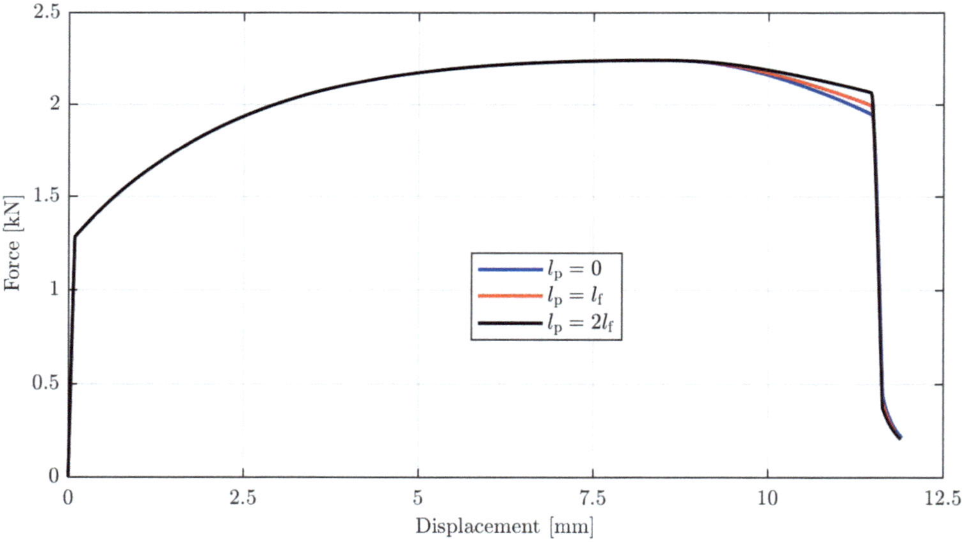

Figure 3.7: **Tensile test.** Load deflection results for different values of l_p.

section in the center of the specimen. This increase in volumetric plastic deformation leads to a shift in crack initiation, which is also shown by the load deflection result in Figure 3.8. Note that the fracture evolution is governed by the elastic response, so the

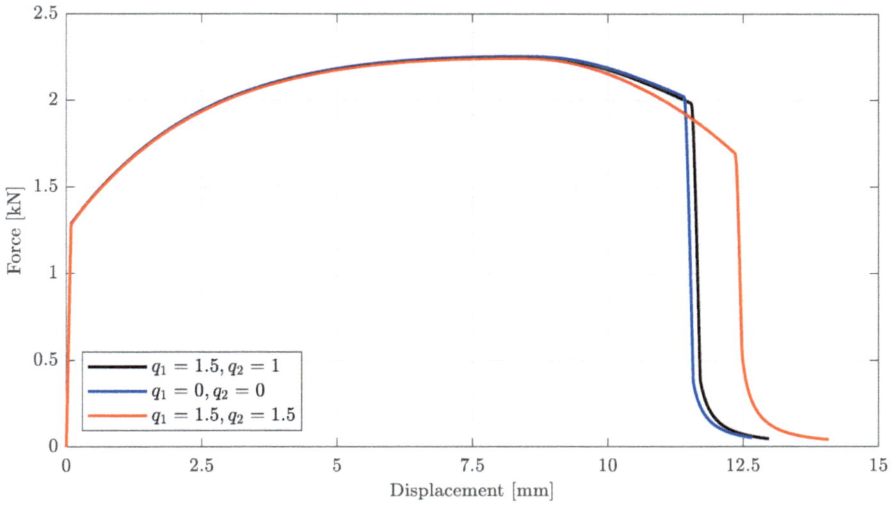

Figure 3.8: **Tensile test.** Load deflection result of isothermal simulations at $\theta = 293\,\mathrm{K}$ and different values of the Gurson fitting parameters.

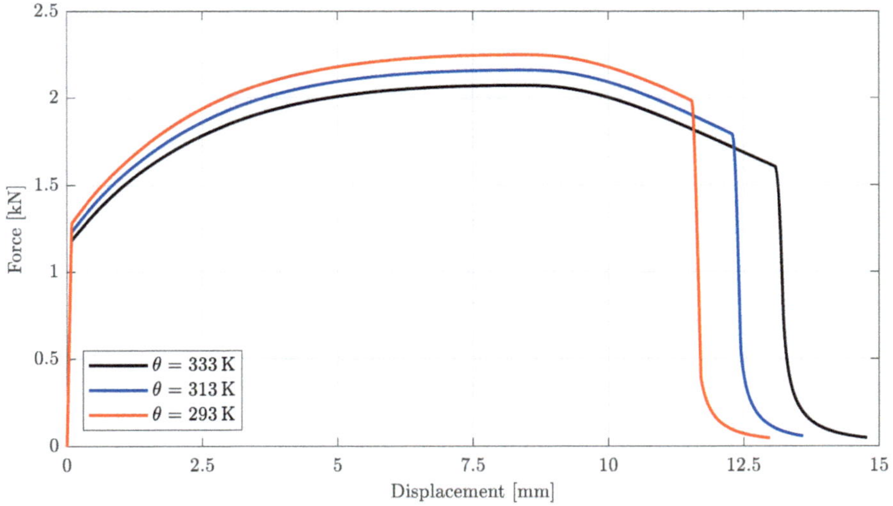

Figure 3.9: **Tensile test.** Load deflection result of isothermal simulations at different temperatures.

maximum stress value at crack initiation is almost identical, but the applied load differs due to the different constricted cross sections.

Concerning **point VIII**, the temperature dependence of the model is addressed next.

Therefore, the load deformation result is depicted in Figure 3.9 for isothermal simulations with different temperatures. As expected, higher temperature reduces the yield stress and due to this effect on the elastic response, the measured displacement at crack initiation is larger, i.e., we obtain larger plastic deformations. This can also be seen in Figure 3.11, where the results of equivalent plastic strain and void volume fraction are plotted.

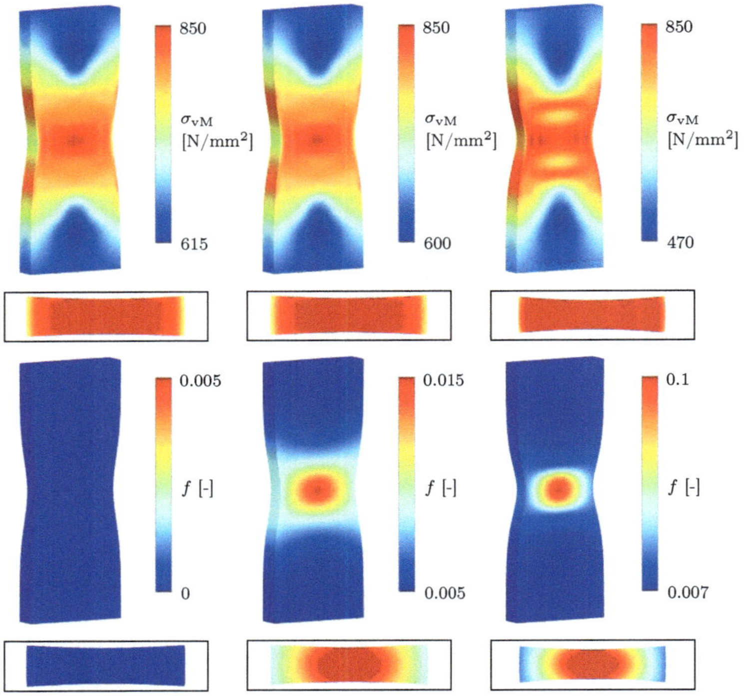

Figure 3.10: **Tensile test.** Result of isothermal simulations at $\theta = 293\,\mathrm{K}$ and different values of Gurson fitting parameters $q_1 = \{0, 1.5, 1.5\}$ and $q_2 = \{0, 1, 1.5\}$ (from left to right). The von Mises stress (first row) and void volume fraction (second row) are plotted at crack initialization.

Adressing **point IX** and **point X**, similar effects can be observed by increasing the deformation rate and dissipation factors for fully coupled simulations with an initial temperature of 293 K. As shown in Figure 3.12, an increase in these parameters leads to greater and more localized heating, and thus greater plastic deformations, as discussed above. However, since the heating is local, unlike the isothermal setting, we obtain a shift in crack initiation to smaller displacements measured at $\pm 25\,\mathrm{mm}$ from the center of the specimen, see Figure 3.14.

Finally, Figure 3.13 shows all relevant physical quantities obtained by the fully coupled simulation with $\dot{u} = 0.5\,\mathrm{mm/min}$ and $\nu_{\mathrm{f}} = \nu_{\mathrm{p}} = 0.9$. To be precise, the crack phase-field, absolute temperature, equivalent plastic strain, dual strain hardening force, void

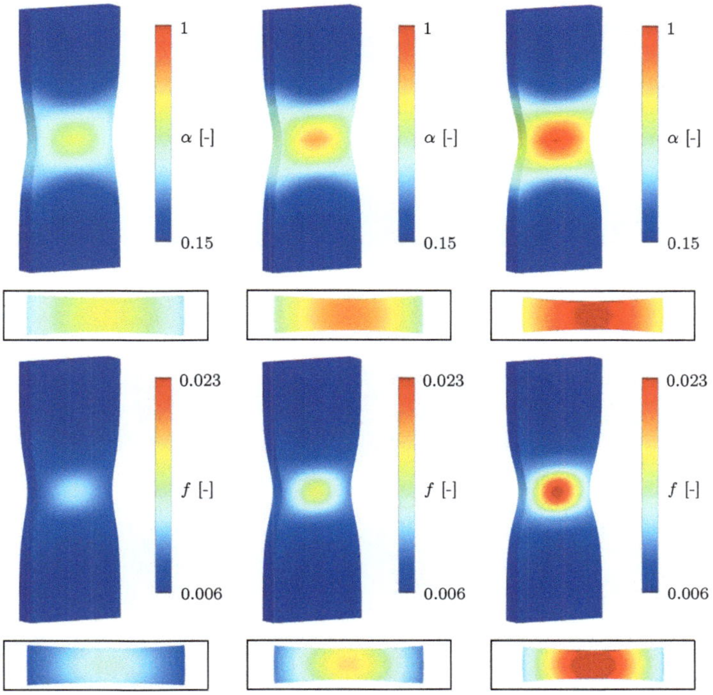

Figure 3.11: **Tensile test.** Result of isothermal simulations at temperatures of $\theta = \{293, 313, 333\}$ K. The equivalent plastic strain (first row) and the void volume fraction (second row) are plotted at crack initialization.

volume fraction, and von Mises stress distribution at different displacement steps are presented. Note that we do not observe any kind of oscillations as addressed in **point VI**. Uniform discretization of all fields, i.e., discretization using quadratic NURBS-based shape functions, leads to strong oscillations in the dual hardening force field and stress field, causing the simulation to stop. It is evident that the plastic deformation remains persistent in case of fracture, so residual stresses remain after the specimen is completely fractured. Also, it is evident from Figure 3.13 that the plastic deformation remains persistent in case of fracture, so residual stresses are generated within the specimen.

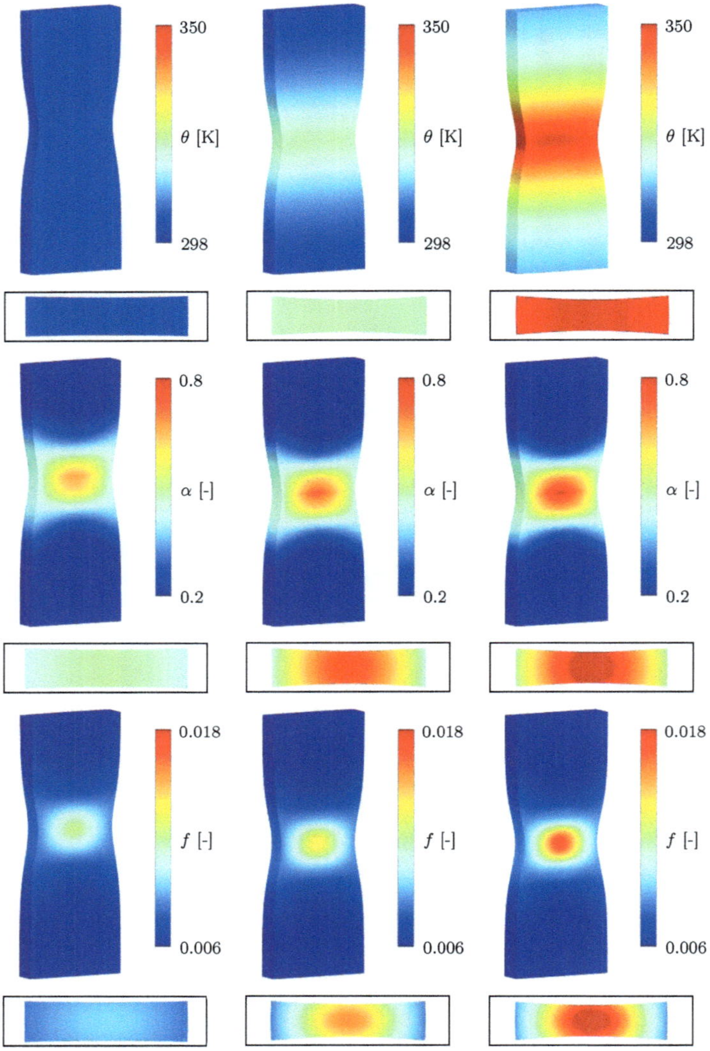

Figure 3.12: **Tensile test.** Absolute temperature result (first row), equivalent plastic strain result (second row) and void volume fraction result (third row) of fully coupled simulations at crack initialization using different deformation rates $\dot{u} = \{0.5, 60, 60\}$ mm/min and dissipation factors $\nu_{\mathrm{f}} = \nu_{\mathrm{p}} = \{0.9, 0.5, 0.9\}$ (from left to right).

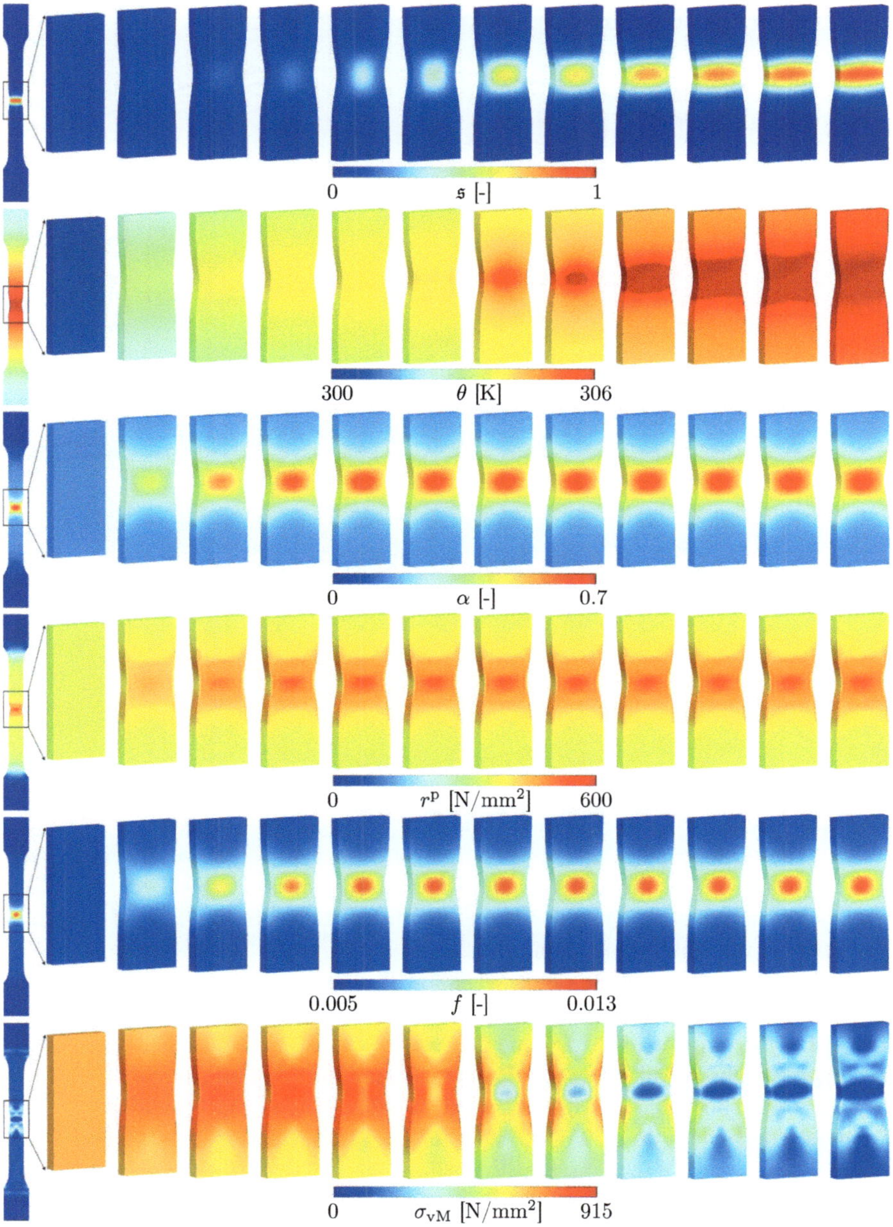

Figure 3.13: **Tensile test.** Crack phase-field result (first row), absolute temperature result (second row), equivalent plastic strain result (third row), dual hardening force result (fourth row), void volume fraction result (fifth row) and von Mises stress result (sixth row) at displacements of $u_{\pm 25}$ = [7.92, 10.12, 10.94, 11.26, 11.46, 11.47, 11.51, 11.52, 11.56, 11.58, 11.59, 11.65] mm.

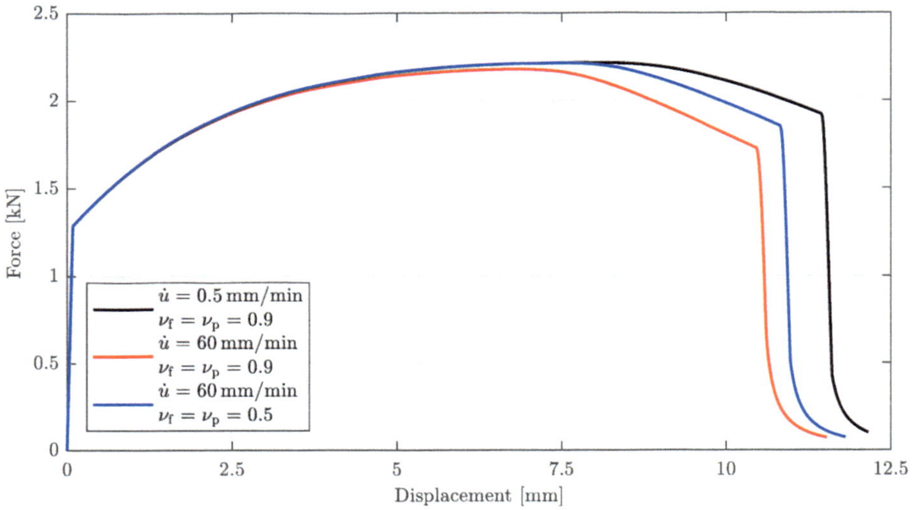

Figure 3.14: **Tensile test.** Load deflection result of fully coupled simulations using different values of the deformation rate and dissipation factors.

3.5.2 Third Sandia Fracture Challenge

This last example illustrates the applicability of the thermo-porous fracture model to complex three-dimensional geometries. Therefore we consider the geometry of the third Sandia Fracture Challenge[†].The geometry in the reference configuration along with the applied boundary conditions and the computational mesh is shown in Figure 3.15. The examined specimen of size $56\,\mathrm{mm} \times 18\,\mathrm{mm} \times 8\,\mathrm{mm}$ with a reduced width of $12\,\mathrm{mm}$ in the middle contains multiple internal channels and cavities. The initial temperature of the body is $293\,\mathrm{K}$ and no heat in- or outflow is allowed. For an efficient and simple construction of the computational mesh based on the given CAD data, the geometry is discretized by 111492 Lagrangian tetrahedral elements, whereby a quadratic approximation is used for the displacement field, the crack phase-field and the temperature field and a linear approximation for both hardening fields. Within the quasi-static simulation setting, the lower end of the hollow body is fixed in space while the upper end is moved upwards by a displacement rate of $0.2\,\mathrm{mm/min}$ Regarding the material setting, steel like parameters are applied as given in Table 3.2 along with plastic and fracture dissipation factors $\nu_\mathrm{p} = \nu_\mathrm{f} = 0.5$ and GTN parameters $f_0 = 0.0005$, $q_1 = 0.75$ and $q_2 = 0.5$.

Figure 3.18 shows the crack phase-field, the equivalent plastic strain, the von Mises stress distribution and the absolute temperature at different displacements steps. The crack

[†] CAD data are taken from the following webpage:
https://drive.google.com/drive/folders/0B8nFwdOwWu0YRkJ4Q0VJWEVCdGM.

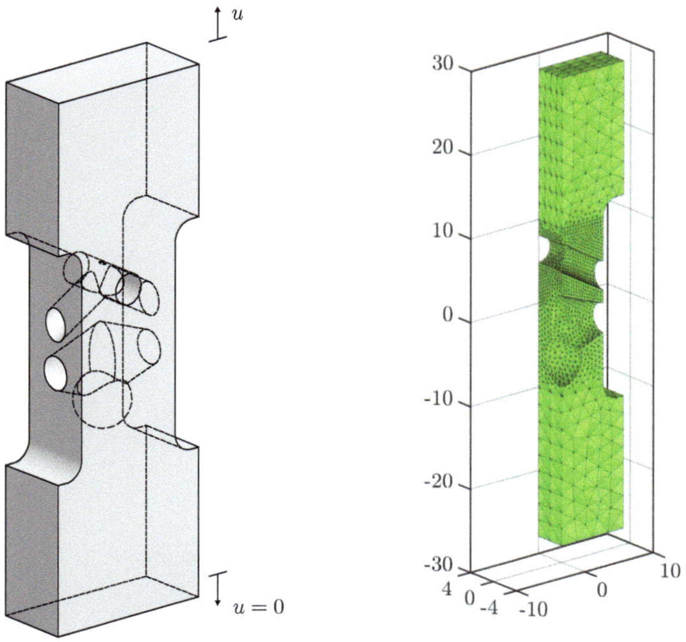

Figure 3.15: **Third SFC.** Reference configuration and computational mesh (sectional view).

phase-field initiates near the intersection between the through hole and the angled channels which is driven by the plastic strain evolution. Subsequently, the crack propagates along the channels as shown in Figure 3.18 (first row). At the fully broken state, plastic deformations remain persistent within the specimen inducing pronounced residual stresses, see Figure 3.18 (third row). Regarding the thermal field, a global cooling down can be observed in the beginning due to the expansion of the specimen. Afterwards, the specimen warms up due to plastification and fracture induced energy dissipation. At the fully ruptured state, an increase of temperature of about 2 K can be observed, see Figure 3.18 (fourth row). Note that this warming affects the hardening behavior and thus also the fracture behavior as pointed out in the previous example. Moreover, the energy dissipation occurs above the conical channel and the internal cavity which impedes an heat flux into the lower part of the specimen such that the upper part warms up slightly faster.

Eventually, the load deflection curve is depicted in Figure 3.16. Therein, we can observe a pronounced hardening and necking behavior before crack initialization. Figure 3.17 shows the results of the crack phase-field at fully ruptured state; the fully broken elements have been removed for visualization purpose.

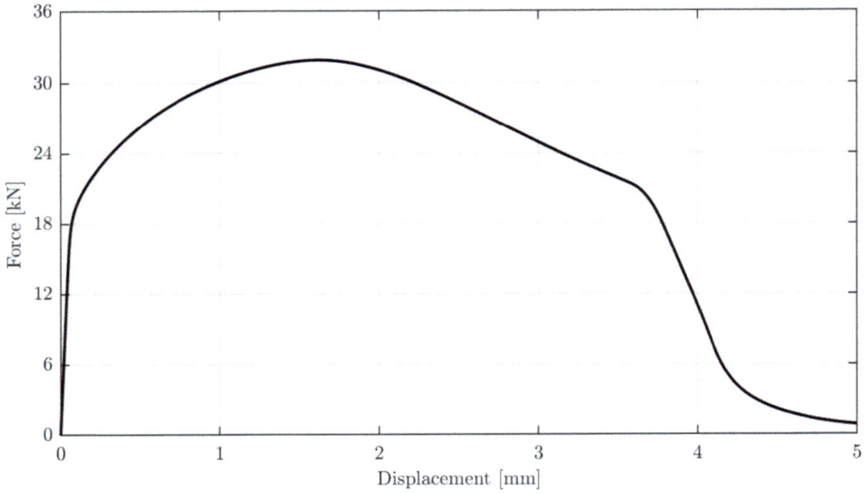

Figure 3.16: **Third SFC.** Load deflection result.

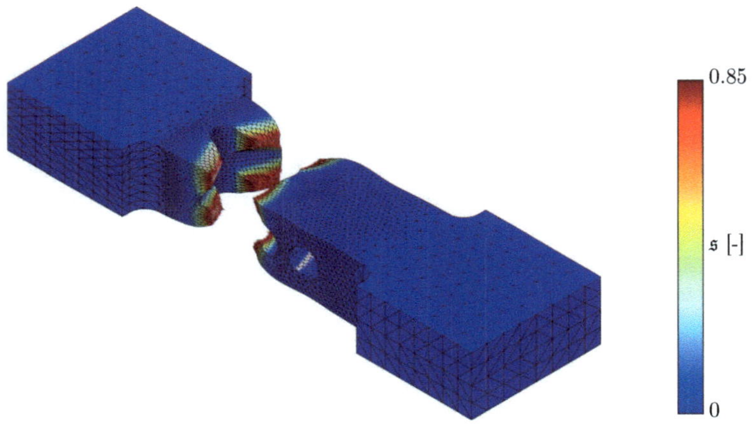

Figure 3.17: **Third SFC.** Results of the crack phase-field at fully ruptured state.

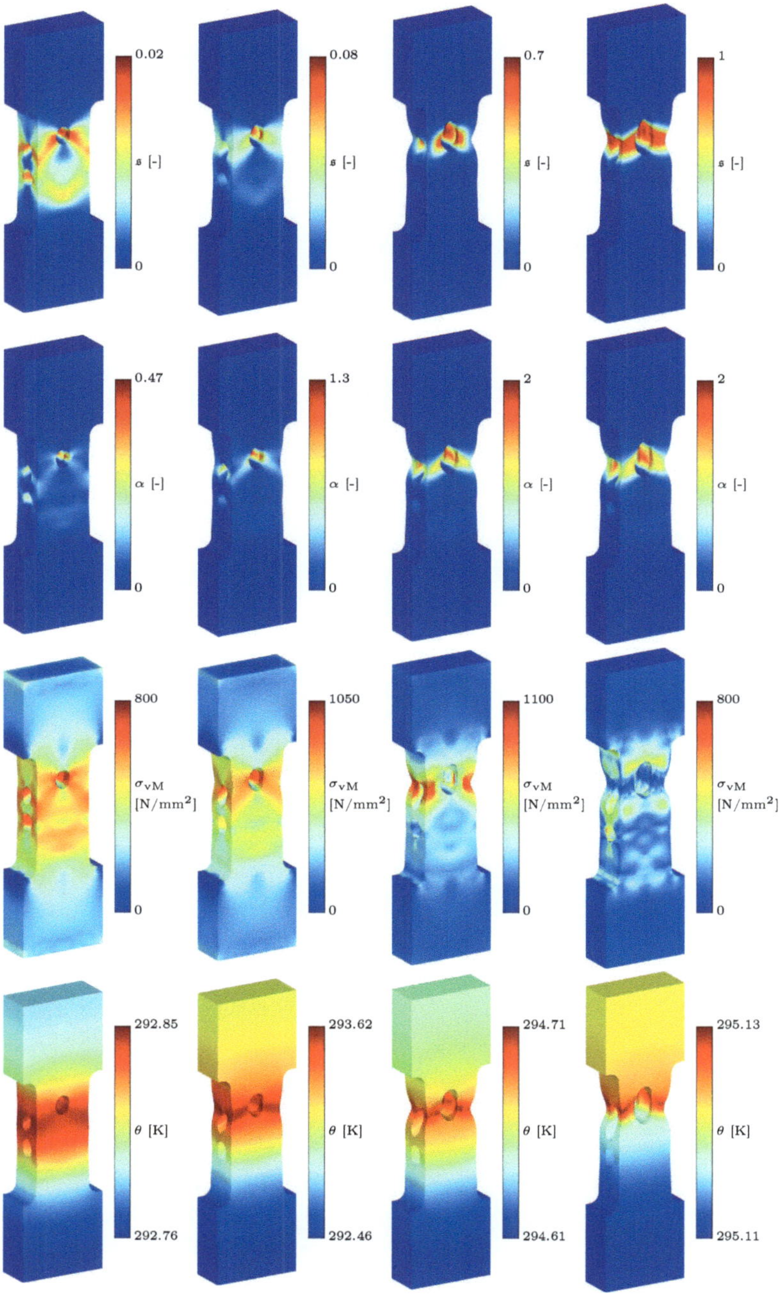

Figure 3.18: **Third SFC.** Crack phase-field result (first row), equivalent plastic strain result (second row), von Mises stress result (third row) and absolute temperature result (fourth row) at displacements of $u = [1.2,\ 2.3,\ 3.9,\ 5]$ mm.

4 Generalized Continuum Mechanics for Micro-Structured Materials

Although most materials look macroscopically homogeneous, on sufficiently small scales they usually encompass a distinct microstructure related to their individual microcomponents, e.g., blood cells, grains, single crystals, fibers, etc. As long as the spatial scales of the continuum differ by many order from those of the underlying microstructure, the classical continuum approach presented in the previous chapter can account for complex material behavior, especially in combination with elaborate constitutive models. However, if a structure under consideration is comparatively small, i.e. approaching a typical length of the intrinsic microstructure, size effects appear, typically characterized by as a stiffer response for a smaller specimen size. In the absence of a characteristic microstructural length scale, e.g. size of embedded fibers or spacing between grains or crystals, classical continuum theory cannot reproduce these macroscopic manifestations of microstructure.

The main objective of this chapter is to introduce the so-called generalized continuum (or microcontinuum) theories to overcome these limitations. Generalized continua can be classified into two main groups, see Figure 4.1 and [81, 80, 50]. Higher grade continua are characterized by higher order spatial derivatives of the displacement field, such as in the second gradient theory of Mindlin [92] or by the gradient of internal variables as in strain-gradient plasticity [1, 99]. Higher order continua, generally known as micromorphic continua, are represented with additional degrees of freedom that are a priori independent from the usual three degrees of freedom of the continuum point [21].

After presenting the basic kinematics following [21], of this two main groups of microstructured continua, we will focus on the micromorphic theory of Eringen and Mindlin [44, 92] for linear elasticity and demonstrate how it incorporates many other models as special cases, such as the Cosserat [25] or second gradient theory [78]. The latter will be used in the following chapters to account for size effects in the modeling of fiber composite reinforcements.

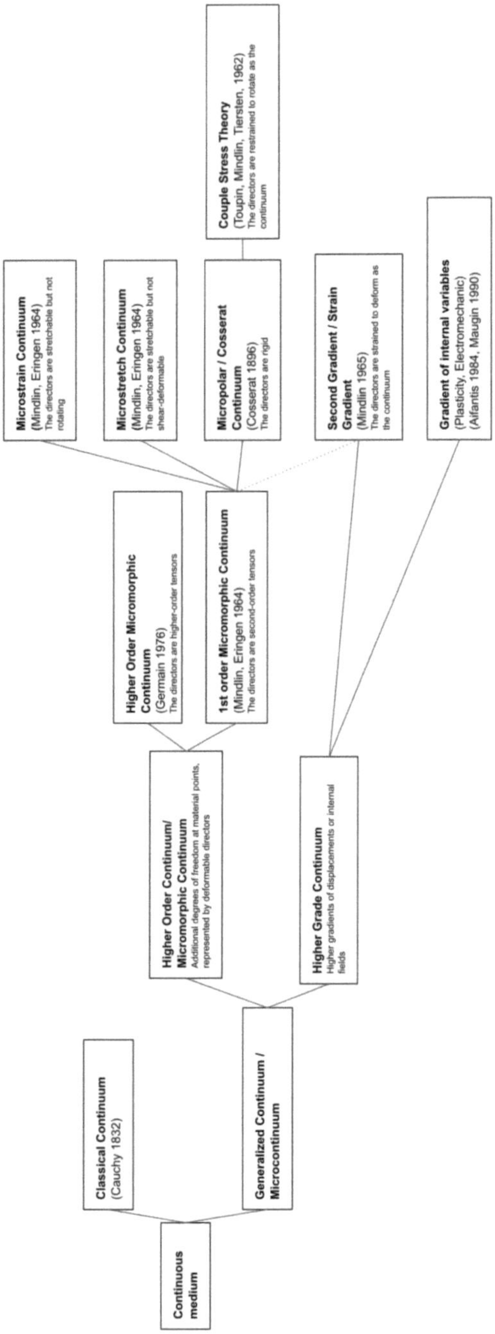

Figure 4.1: Classification of continuum materials.

4.1 Kinematics

4.1.1 Configuration and motions

Higher-order gradient theory

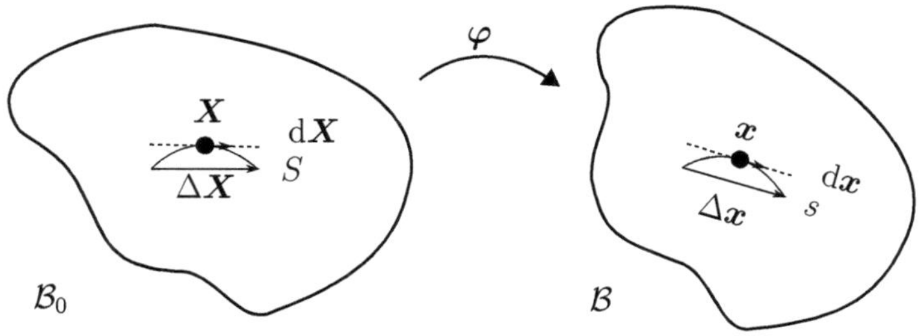

Figure 4.2: Nonlinear deformation mapping

Considering a three-dimensional body in its Lagrangian and Eulerian configuration $\mathcal{B}_0$ and $\mathcal{B}$, the deformation of a material point P, identified by its position vector $\boldsymbol{X}$ and $\boldsymbol{x}$, is described via the mapping

$$\boldsymbol{x} := \boldsymbol{\varphi}(\boldsymbol{X}, t). \tag{4.1}$$

We additionally introduce the deformation of a material line L. The material tangent at the material finite line element S around the material point P with material placement $\boldsymbol{X}$ is denoted by $\Delta\boldsymbol{X}$, as depicted in Figure 4.2. In the vicinity of P, the following Taylor series expansion may be used to generate the mapping of this finite tangent element

$$\Delta\boldsymbol{x} := \nabla\boldsymbol{\varphi} \cdot \Delta\boldsymbol{X} + \frac{1}{2}\nabla(\nabla\boldsymbol{\varphi}) : (\Delta\boldsymbol{X} \otimes \Delta\boldsymbol{X}) + \phi(\Delta\boldsymbol{X}^{3+}). \tag{4.2}$$

In the classical Cauchy-Boltzmann continuum, only the linear tangent map $\nabla\boldsymbol{\varphi}$ is taken into account, and thus neglecting the quadratic term $\nabla(\nabla\boldsymbol{\varphi})$ and terms of cubic and higher degree $\phi(\delta\boldsymbol{X}^{3+})$. This requires that the dimensions over which macroscopic gradients occur are much larger than characteristic length scales of the microstructure. As a consequence, only infinitesimal line elements or the tangent vector $\mathrm{d}\boldsymbol{X}$ on S at the position $\boldsymbol{X}$ affects the deformation map.

$$\mathrm{d}\boldsymbol{x} := \nabla\boldsymbol{\varphi} \cdot \mathrm{d}\boldsymbol{X}. \tag{4.3}$$

When the linear term in the nonlinear deformation mapping (4.2) is combined with the quadratic or even higher terms, the classical continuum is extended to a gradient continuum. When third- and higher-order terms are omitted, this corresponds to a second-order gradient theory, which is a specific instance of Mindlin's second gradient model [91]. As will be explained in more detail later, a size effect is introduced by the second-gradient $\nabla(\nabla\boldsymbol{\varphi})$, which can be related to the curvature in the material line and thus will be used to model the size-dependent curvature effects of the embedded fibers.

Higher gradients of internal fields are used analogously e.g. in theories of gradient plasticity or gradient electromechanics, see [41, 1, 99, 100, 101].

Micromorphic media

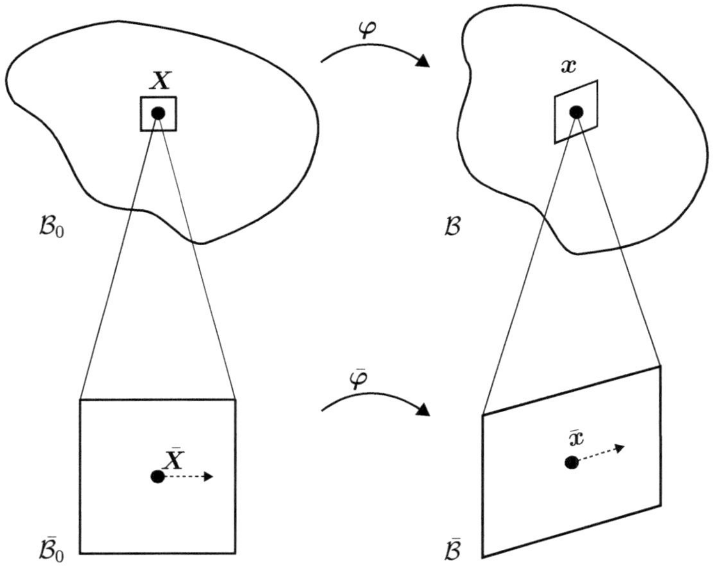

Figure 4.3: Micromorphic deformation maps.

Rather than considering higher gradients of deformation, the micromorphic theory extends the classical continuum in a more general approach by introducing additional degrees of freedom per continuum point. More precisely, at each point P in the (macro)continuum, a microcontinuum $\bar{\mathcal{B}}_0$ with a set of directors $\bar{\boldsymbol{x}}_\alpha$ with $\alpha = 1, ..., N$ is attached, see Figure 4.3. These directors represent additional degrees of freedom required to characterize the microstructural behavior, and denote the orientations and intrinsic deformations of the material points of P. Particularly well demonstrated in the physical representations of microdeformation in crystalline solids [21], both the geometric point P and the vectors $\bar{\boldsymbol{x}}_\alpha$ have their own independent motion. In contrast to highe- order micromorphic continua where the directors are higher-order tensors, we focus on first-order micromorphic continua with $N = 1$ and second-order directors.

In addition to the classical deformation mapping, the so called macromotion

$$\boldsymbol{x} := \boldsymbol{\varphi}(\boldsymbol{X}, t) \tag{4.4}$$

we define a corresponding mapping for the micromotion, mapping the attached directors from the reference configuration into the current one

$$\bar{\boldsymbol{x}} := \hat{\varphi}(\boldsymbol{X}, \bar{\boldsymbol{X}}, t). \tag{4.5}$$

In comparison to the macroscopic scales of the body, the material particles are considered to be of very small size. Consequently, the micromotion 4.5 is generally linearly approximated as

$$\bar{\boldsymbol{x}} := \bar{\varphi}(\boldsymbol{X}, t) \cdot \bar{\boldsymbol{X}} \tag{4.6}$$

with the second-order microdeformation tensor $\bar{\varphi}(\boldsymbol{X}, t)$.

As originally defined by Eringen [44], a material body is referred to as a micromorphic continuum (of grade one), if its motions are described by (4.4) and (4.6) and have unique inverses

$$\begin{aligned} \boldsymbol{X} &= \varphi^{-1}(\boldsymbol{X}, t) \\ \bar{\boldsymbol{X}} &= \bar{\varphi}^{-1}(\boldsymbol{X}, t) \cdot \bar{\boldsymbol{x}}. \end{aligned} \tag{4.7}$$

In order to retain the right-hand screw orientations of the frame of reference, and ensure the physical assumption of continuity, indestructibility and impenetrability of matter, we assume

$$\det(\nabla \varphi) > 0 \tag{4.8}$$

and

$$\det(\bar{\varphi}) = 1/\det(\bar{\varphi}^{-1}) > 0. \tag{4.9}$$

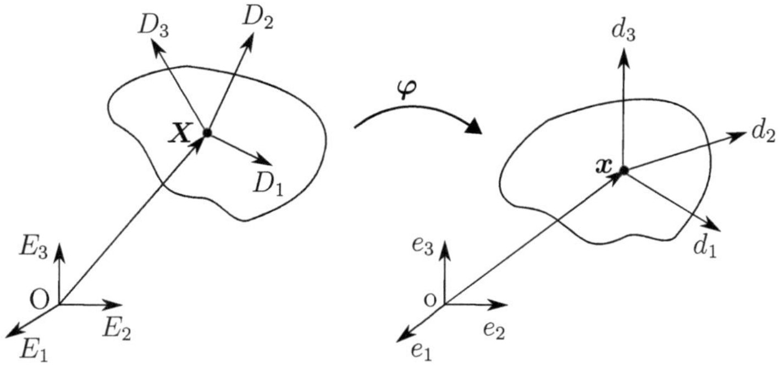

Figure 4.4: Deformable directors of the micromorphic continuum

As shown in Figure 4.4, a material point possesses, in addition to the three usual translations, three deformable directors which represent the degrees of freedom arising from microdeformations and are denoted by

$$\begin{aligned} d_j &= \bar{\varphi}_{ij}(\boldsymbol{X}, t) e_i, \\ D_j &= \bar{\varphi_{ij}}^{-1}(\boldsymbol{X}, t) E_i \end{aligned} \tag{4.10}$$

with the components cartesian unit vectors E_i and $e_i (i = 1, 2, 3)$ in the material and spatial configuration, respectively.

4.1.2 Rotation and subclasses

With the three distinct deformable directors, and its total of 12 degrees of freedom, the general micromorphic continuum formulation is appropriate for the modelling of many complex materials like polymers with flexible molecules, liquid crystals with side chains, animal blood with deformable cells, etc. Its microdeformation can be illustrated with the deformable director triad shown in Figure 4.5.

Moreover, the micromorphic continuum includes many subtheories, each of which is defined by imposing specific constraints on the microdeformation, see Figure 4.1 . A detailed systematic presentation can be found in [50]. In what follows, we will briefly discuss the most common of these, namely the micropolar, the mircrostretch and the microstrain continuum. Later in the chapter, we will show how a second-gradient theory can be obtained by imposing analogous kinematic constraints on the generalized micromorphic continua.

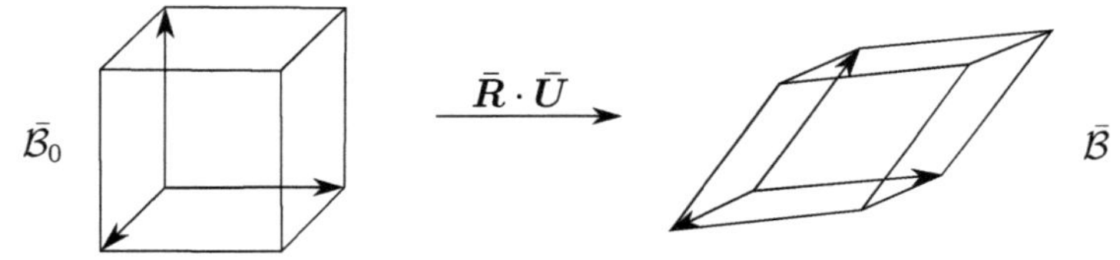

Figure 4.5: Micromorphic deformation (deformable director triad)

Just like the deformation gradient introduced in Chapter 2, the microdeformation tensor includes both stretch and rotation

$$\bar{\varphi} = \bar{\boldsymbol{R}} \cdot \bar{\boldsymbol{U}}, \tag{4.11}$$

where $\bar{\boldsymbol{R}}$ is the orthogonal microrotation tensor (i.e., $\bar{\boldsymbol{R}}^{-1} = \bar{\boldsymbol{R}}^T$ and $\det(\bar{\boldsymbol{R}}) = 1$) and $\bar{\boldsymbol{U}}$ is the symmetric and definite positive right microstretch tensor.

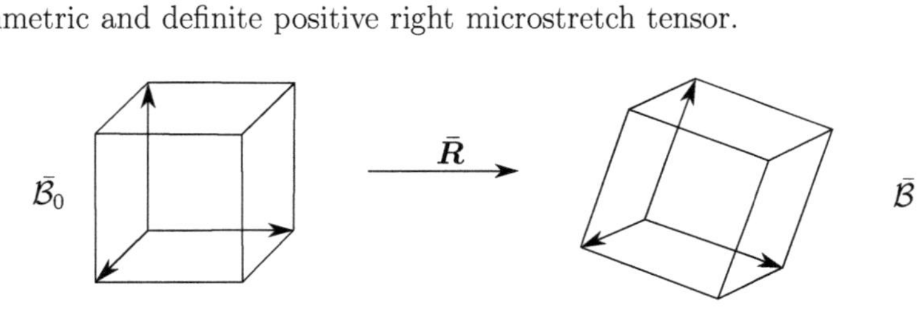

Figure 4.6: Micropolar deformation (rigid director triad)

A **micropolar continuum** (often simply referred to as Cosserat continuum) as illustrated with the rigid director triad in Figure 4.6 may only experience rotation

$$\bar{\boldsymbol{x}} = \bar{\boldsymbol{R}} \cdot \bar{\boldsymbol{X}}. \tag{4.12}$$

The micropolar theory is typically used for the investigation of granular or multimolecular materials, like rigid chopped fibres, elastic solids with rigid granular inclusions, liquid crystals, animal blood, etc. [15, 120].

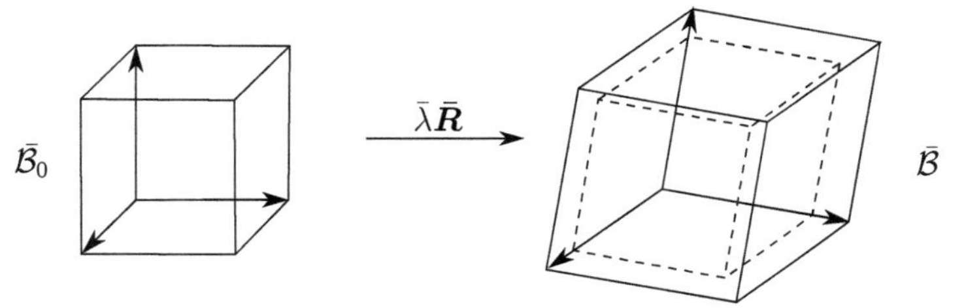

Figure 4.7: Microstretch deformation (extensible director triad)

Besides a rotation $\bar{\boldsymbol{R}}$, also one scalar stretch variable $\bar{\lambda}$ is considered to account for isotropic expansion and contraction of a **microstretch continuum** without microshearing

$$\bar{\boldsymbol{x}} = \bar{\lambda}\bar{\boldsymbol{R}} \cdot \bar{\boldsymbol{X}}. \tag{4.13}$$

In addition to the micropolar director triad with fixed angles, the director triad of is isotropically deformable as shown in Figure 4.7. Chopped elastic fibres, animal bones or inviscid liquids are typically simulated as a microstrech continuum [50].

In a **microstrain continuum**, the rotation is neglected, rather only the stretch tensor $\bar{\boldsymbol{U}}$ is considered

$$\bar{\boldsymbol{x}} = \bar{\boldsymbol{U}} \cdot \bar{\boldsymbol{X}}, \tag{4.14}$$

as illustrated in Figure 4.8 . The director triad allows for pure stretch, which is characterized by the fact that the principle directions maintain their direction. The microstrain theory is used e.g. for metallic foams [50].

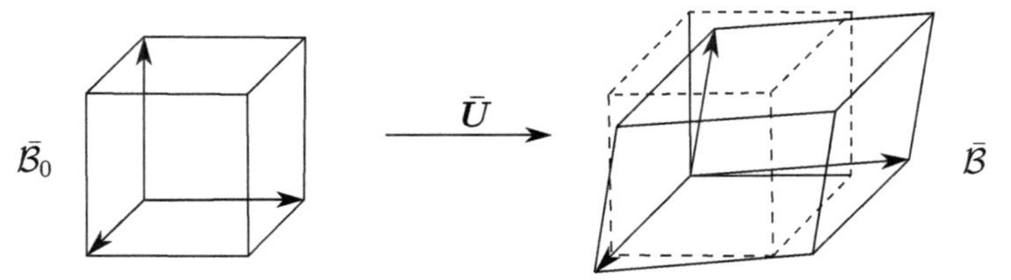

Figure 4.8: Microstrain deformation (stretchable director triad excluding rotation)

4.2 Micromorphic linear elasticity

With the introduced kinematic framework for generalized continuum mechanics, a mature and growing field of research has emerged over the last 50 years to conceptualize appropriate constitutive theories and corresponding deformation measures for a wide range of microstructured materials. In the following chapters, we will present a specific microstructured constitutive formulation at finite strains for the mechanical simulation of fiber composite reinforcements. However, since the main objective of this chapter is to give a concise introduction to general microcontinuum theories and demonstrate how they can be suitably constrained to obtain more specific models, i.e., a second-gradient theory, we next present the linear theory of micromorphic elasticity developed by Mindlin [92]. Moreover, the paradigmatic structure of the linear model allows a more straightforward understanding of the general concept, by avoiding the mathematical complexity brought by nonlinearities. This specific linear model has been used to describe a variety of physical phenomena, such as polymers with deformable molecules, biological tissues, simple harmonic waves, etc. [43].

4.2.1 Deformation measures

Rather than the (macroscopic) deformation map $\boldsymbol{\varphi}$, we consider the classical displacement field in the small strain theory

$$\boldsymbol{u} = \boldsymbol{x} - \boldsymbol{X}. \tag{4.15}$$

Moreover, Mindlin specified the following objective Lagrangian strain measures for infinitesimal deformations, namely the classical linearized macro deformation strain tensor

$$\boldsymbol{E} = \frac{1}{2}(\nabla\boldsymbol{u} + \nabla\boldsymbol{u}^{\mathrm{T}}), \tag{4.16}$$

the relative micro-macro deformation

$$\boldsymbol{\gamma} = \nabla\boldsymbol{u} - \bar{\boldsymbol{\varphi}}, \tag{4.17}$$

and the gradient of the microdeformation (third-order tensor)

$$\boldsymbol{\mathfrak{K}} = \nabla\bar{\boldsymbol{\varphi}}. \tag{4.18}$$

4.2.2 Energetic response

The general form of the micromorphic strain energy density introduced in [92] is a function of the 42 kinematic variables E_{ij}, γ_{ij} and $\mathfrak{K}_{ijk}$, and is of the type

$$\begin{aligned}\Psi = \Psi(E_{ij}, \gamma_{ij}, \mathfrak{K}_{ijk}) =& \frac{1}{2}a_{ijkl}E_{ij}E_{kl} + \frac{1}{2}b_{ijkl}\gamma_{ij}\gamma_{kl} + \frac{1}{2}c_{ijklmn}\mathfrak{K}_{ijk}\mathfrak{K}_{lmn} \\ &+ d_{ijklm}\gamma_{ij}\mathfrak{K}_{klm} + e_{ijklm}\mathfrak{K}_{ijk}E_{lm} + f_{ijkl}\gamma_{ij}E_{kl}.\end{aligned} \tag{4.19}$$

Due to the symmetries of the strain tensors, only 903 of the 1764 introduced scalar material coefficients are independed. Assuming an isotropic material, the number of independent constitutive parameters further decreases to 18 (see [92]), leading to an energy function

$$\begin{aligned}\Psi =&\frac{1}{2}\lambda E_{ii}E_{jj} + \mu E_{ij}E_{ij} + \frac{1}{2}b_1\gamma_{ii}\gamma_{jj} + \frac{1}{2}b_2\gamma_{ij}\gamma_{ij} + \frac{1}{2}b_3\gamma_{ij}\gamma_{ji}\\ &+ c_1\mathfrak{K}_{iik}\mathfrak{K}_{kjj} + c_2\mathfrak{K}_{iik}\mathfrak{K}_{jkj} + \frac{1}{2}c_3\mathfrak{K}_{iik}\mathfrak{K}_{jjk} + \frac{1}{2}c_4\mathfrak{K}_{ijj}\mathfrak{K}_{ikk}\\ &+ c_5\mathfrak{K}_{ijj}\mathfrak{K}_{kik} + \frac{1}{2}c_6\mathfrak{K}_{iji}\mathfrak{K}_{kjk} + \frac{1}{2}c_7\mathfrak{K}_{ijk}\mathfrak{K}_{ijk} + c_8\mathfrak{K}_{ijk}\mathfrak{K}_{jki}\\ &+ \frac{1}{2}c_9\mathfrak{K}_{ijk}\mathfrak{K}_{ikj} + \frac{1}{2}c_{10}\mathfrak{K}_{ijk}\mathfrak{K}_{jik} + \frac{1}{2}c_{11}\mathfrak{K}_{ijk}\mathfrak{K}_{kji}\\ &+ f_1\gamma_{ii}E_{jj} + f_2(\gamma_{ij}+\gamma_{ji})E_{ij}.\end{aligned} \tag{4.20}$$

Depending on the particular microstructured material behavior under consideration, more specific and simplified formulations have been developed. For instance, an energy density function with only 6 constitutive coefficients was introduced in [78], which qualitatively still reflects all material characteristics of the general micromorphic model of Mindlin

$$\begin{aligned}\Psi = \mu_e||\mathrm{sym}(\nabla\boldsymbol{u} - \bar{\boldsymbol{\varphi}})||^2 + \frac{\lambda_e}{2}(\mathrm{tr}(\nabla\boldsymbol{u} - \bar{\boldsymbol{\varphi}}))^2 + \mu_h||\mathrm{sym}(\bar{\boldsymbol{\varphi}})||^2 + \frac{\lambda_h}{2}(\mathrm{tr}(\bar{\boldsymbol{\varphi}}))^2\\ \mu_c||\mathrm{skew}(\nabla\boldsymbol{u} - \bar{\boldsymbol{\varphi}})||^2 + \frac{\alpha_g}{2}||\nabla\bar{\boldsymbol{\varphi}}||^2.\end{aligned} \tag{4.21}$$

Here, the microscopic Lamé moduli μ_h and λ_h are related to the response of a representative volume element of the substructure. Along with the classical macroscopic Lamé moduli μ and λ obtained in experiments for large material samples with assumed heterogeneous structure and a characteristic length of zero, the isotropic scale transition parameters μ_e and λ_e can be determined from homogenization theory, see [78]. The Cosserat couple modulus μ_c which governs the asymmetry of the force stresses is i.e. used to describe the mechanical behavior of very particular metamaterials as lattice structures and phonon crystals. Finally, α_g accounts for different microstructural effects, i.e. the curvature depencence or some particular optic waves propagation.

Along with the following restrictions for the introduced material parameters

$$\mu_e > 0, \quad \mu_c > 0, \quad 3\lambda_e + 2\mu_e > 0, \mu_h > 0, \quad 3\lambda_h + 2\mu_h > 0, \quad \alpha_c > 0. \tag{4.22}$$

the well-posedness and positive definiteness of the formulation has been outlined in [78]. A thorough identification of the relations for the 6 coefficients proposed here to specific micro and macro deformation modes, and the relations in terms of the parameters introduced by Mindlin is outlined in [78].

Although we will ignore inertia effects in the scope of this work, we briefly introduce the micromorphic kinetic energy potential for the sake of completeness, as

$$T = \frac{1}{2}\rho\,\|\boldsymbol{u}_{,t}\|^2 + \frac{1}{2}\bar{\rho}\,\|\bar{\boldsymbol{\varphi}}_{,t}\|^2 \tag{4.23}$$

with the macroscopic and microscopic mass densities ρ and $\bar{\rho}$ defined per unit of macro volume.

Considering the dimensionless of the second order tensor $\bar{\boldsymbol{\varphi}}$, the coefficient $\bar{\rho}$ has a dimension of a bulk density multiplied by the square of the length. With the real density of the microstructure ρ' , and a characteristic length l directly associated to the characteristic size of the microscopic inclusions, the microscopic mass density may be written as

$$\bar{\rho} = l^2 \rho', \tag{4.24}$$

see [78].

Note that in contrast to this special form with a single characteristic length, there are also more general types of kinetic energies that take into account different characteristic lengths l_{ij} and therefore for more intricate microstructures [92].

4.2.3 Strong form

Ignoring inertia effects, the equations of motion for a linear micromorphic continuum will be derived through the principle of virtual work $\delta W^{\text{int}} - \delta W^{\text{ext}} = 0$.

For the elastic potential energy given in (4.21), we derive the internal virtual work as

$$\delta W^{\text{int}} = \int_{B_0} \left(\frac{\partial \Psi}{\partial \boldsymbol{E}} : \delta \boldsymbol{E} + \frac{\partial \Psi}{\partial \gamma} : \delta \gamma + \frac{\partial \Psi}{\partial \mathfrak{K}} :\cdot \delta \mathfrak{K} \right) \mathrm{d}V. \tag{4.25}$$

Regarding the partial derivatives therein, we introduce the work conjugate stresses to the strain components

$$\begin{aligned} \boldsymbol{\sigma}_1 &= \frac{\partial \Psi}{\partial \boldsymbol{E}} = \boldsymbol{\sigma}_1^T, \\ \boldsymbol{\sigma}_2 &= \frac{\partial \Psi}{\partial \gamma}, \\ \mathfrak{S} &= \frac{\partial \Psi}{\partial \mathfrak{K}}, \end{aligned} \tag{4.26}$$

namely, the Cauchy stress, the relative stress and the double stress, respectively.

Using the symmetry of $\boldsymbol{\sigma}_1$ and the definition of the corresponding deformation measures, we obtain

$$\delta W^{\text{int}} = \int_{B_0} \boldsymbol{\sigma}_1 : \delta \nabla \boldsymbol{u} + \boldsymbol{\sigma}_2 : (\delta \nabla \boldsymbol{u} - \delta \bar{\varphi}) + \mathfrak{S} :\cdot \delta \nabla \bar{\varphi} \, \mathrm{d}V \tag{4.27}$$

After an integration by parts

$$\delta W^{\text{int}} = \int_{B_0} -\nabla\cdot(\boldsymbol{\sigma}_1+\boldsymbol{\sigma}_2)\cdot\delta\boldsymbol{u}+\nabla\cdot(\delta\boldsymbol{u}\cdot(\boldsymbol{\sigma}_1+\boldsymbol{\sigma}_2))-(\boldsymbol{\sigma}_2+\nabla\cdot\mathfrak{S})\cdot\delta\bar{\varphi}+\nabla\cdot(\delta\bar{\varphi} : \mathfrak{S}) \, \mathrm{d}V \tag{4.28}$$

the application of the divergence theorem with the unit normal vector $\boldsymbol{N}$ to the boundary surface ∂B_0 yields

$$\delta W^{\text{int}} = \int_{B_0} -\nabla\cdot(\boldsymbol{\sigma}_1+\boldsymbol{\sigma}_2)\cdot\delta\boldsymbol{u}-(\boldsymbol{\sigma}_2+\nabla\cdot\mathfrak{S})\cdot\delta\bar{\boldsymbol{\varphi}}\,\mathrm{d}V+\int_{\partial B_0}((\boldsymbol{\sigma}_1+\boldsymbol{\sigma}_2)\boldsymbol{N})\cdot\delta\boldsymbol{u}+\mathfrak{S}\boldsymbol{N}:\delta\bar{\boldsymbol{\varphi}}\,\mathrm{d}A. \tag{4.29}$$

Note that all further integrations by parts incorporating the boundaries $\partial^2 B_0$ and $\partial^3 B_0$ and subsequently leading to external edge and wedge forces are omitted for the sake of conciseness.

Following [78] the external contributions can be formulated as

$$\delta W^{\text{ext}} = \int_{B_0} \boldsymbol{b}\cdot\delta\boldsymbol{u}+\boldsymbol{B}:\delta\bar{\boldsymbol{\varphi}}\,\mathrm{d}V+\int_{\Gamma_0^t}\boldsymbol{t}\cdot\delta\boldsymbol{u}\,\mathrm{d}A+\int_{\Gamma_0^T}\boldsymbol{T}:\delta\bar{\boldsymbol{\varphi}}\,\mathrm{d}A \tag{4.30}$$

where $\boldsymbol{b}$ is an external body force per unit volume, $\boldsymbol{B}$ an external double force per unit volume, $\boldsymbol{t}$ is an external traction force per unit area and $\boldsymbol{T}$ is an external double traction force per unit area, acting at the respective Neumann boundaries Γ_0^t and Γ_0^T (see ? for further details).

Eventually, assuming arbitrary variations of the introduced kinematical fields $\delta\boldsymbol{u}$ and $\delta\bar{\boldsymbol{\varphi}}$, we obtain the local form of the problem as

$$\begin{aligned}\nabla\cdot(\boldsymbol{\sigma}_1+\boldsymbol{\sigma}_2)+\boldsymbol{b}&=\boldsymbol{0}\\ \nabla\cdot\mathfrak{S}+\boldsymbol{\sigma}_2+\boldsymbol{B}&=0\end{aligned} \tag{4.31}$$

supplemented by the boundary conditions

$$\begin{aligned}\boldsymbol{u}=\boldsymbol{u}_D &\quad \text{on} \quad \Gamma_0^{\boldsymbol{u}}\\ \bar{\boldsymbol{\varphi}}=\bar{\boldsymbol{\varphi}}_D &\quad \text{on} \quad \Gamma_0^{\bar{\boldsymbol{\varphi}}}\\ (\boldsymbol{\sigma}_1+\boldsymbol{\sigma}_2)\boldsymbol{N}=\boldsymbol{t} &\quad \text{on} \quad \Gamma_0^t\\ \mathfrak{R}\boldsymbol{N}=\boldsymbol{T} &\quad \text{on} \quad \Gamma_0^T\end{aligned} \tag{4.32}$$

with prescribed fields $\boldsymbol{u}_D$ and $\bar{\boldsymbol{\varphi}}_D$, at the mechanical Dirichlet boundaries $\Gamma_0^{\boldsymbol{u}}$ and $\Gamma_0^{\bar{\boldsymbol{\varphi}}}$.

4.2.4 Special case: second-gradient / strain-gradient theory

We have already shown in Section 4.1.2, it is possible to impose appropriate constraints on the additional degrees of freedom inherent in a micromorphic theory in order to derive more specialized models, such as the micropolar/Cosserat theory. In this final section, we will derive the second-gradient model from the micromorphic theory for linear elasticity introduced earlier.

In particular, we have to constrain any relative micro-macro deformation

$$\boldsymbol{\gamma} = \nabla \boldsymbol{u} - \bar{\boldsymbol{\varphi}} \stackrel{!}{=} \mathbf{0} \tag{4.33}$$

and subsequently enforcing the microdeformation to be equal to the gradient of (macro) displacement

$$\bar{\boldsymbol{\varphi}} \stackrel{!}{=} \nabla \boldsymbol{u}. \tag{4.34}$$

The double stress then reads

$$\mathfrak{S} = \frac{\partial \Psi}{\partial \nabla \nabla \boldsymbol{u}}, \tag{4.35}$$

leading to the internal virtual work

$$\delta W^{\text{int}} = \int_{B_0} \boldsymbol{\sigma}_1 : \delta \nabla \boldsymbol{u} + \mathfrak{S} \mathbin{\vdots} \delta \nabla \nabla \boldsymbol{u} \, \mathrm{d}V. \tag{4.36}$$

and the equilibrium condition

$$\nabla \cdot (\boldsymbol{\sigma}_1 - \nabla \cdot \mathfrak{S}) + \boldsymbol{b} = \mathbf{0}. \tag{4.37}$$

Regarding the simplified strain energy function (4.21), this limit case can be numerically obtained e.g. by simultaneously enforcing $\mu_e \to \infty$ and $\mu_c \to \infty$ and thus obtaining a function of the first- and second-gradient of the displacement

$$\Psi(\nabla \boldsymbol{u}, \nabla \nabla \boldsymbol{u}) = \mu_h \left\| \operatorname{sym}(\nabla \boldsymbol{u}) \right\|^2 + \frac{\lambda_h}{2} (\operatorname{tr}(\nabla \boldsymbol{u}))^2 + \frac{\alpha_g}{2} \left\| \nabla \nabla \boldsymbol{u} \right\|^2 . \tag{4.38}$$

Note that alternative equivalent formulations can be derived with the strain tensor and its gradient $\Psi((\boldsymbol{E}, \nabla \boldsymbol{E})$ or analogously for finite strains with the deformation gradient and its gradient $\Psi((\boldsymbol{F}, \nabla \boldsymbol{F})$ as shown in the next chapter, thereupon "strain-gradient" and "second-gradient" theories are often used interchangeably.

5 A Non-Linear Strain-Gradient Formulation for Fiber-Reinforced Composites

In this chapter we focus on a generalized continuum formulation with higher-order contributions for mechanical simulation of fiber materials. In the modeling of fiber-reinforced composites, it is well established to consider the fiber direction in the stored energy in order to account for the transverse isotropy of the overall material, induced by a single family of fibers. However, this approach does not account for the length scale dependent size effects, i.e. the in-plane flexural resistance of the fibers. By using a generalized continuum model based on the strain-gradient, the gradient of the fiber direction vector can be considered as an additional parameter of the stored energy density function. As a result, the improved model presented in this chapter accounts for the bending stiffness of the fibers, thereby allowing independent material modeling without recalibration for the specific fiber direction or load case. Along with additional material parameters, increased continuity requirements on the weak Sobolev space follow in the finite element analysis. Here, the isogeometric approach offers a framework that satisfies these weak formulation criteria.

The chapter is structured as follows: Firstly, a second-gradient theory is applied to Kirchhoff-Love shell elements. In particular, we embed a model of a woven fabric as presented in the work of Steigmann [116] to account for the in- and out-of-plane flexural resistances of the fibers. Shells, as dimensionally reduced approximations of three-dimensional continua, enable a more straightforward presentation including a micromechanical and empirical motivation as well as a detailed verification of the higher-gradient contributions with analytical results. Moreover, the higher-gradient formulation of the fabric is validated with experimental measurements on organic sheets. Finally, a corresponding formulation for fiber-reinforced materials will be extended to general three-dimensional generalized continua, and verified by the means of two bending tests.

Note that throughout the chapter plastic, damage, fracture, and thermal effects as well as fiber-matrix interactions are omitted for now and considered later.

5.1 Shell element formulation

5.1.1 Micromechanical and experimental motivation

The classical constitutive equations are homogeneous and do not include any natural length scale, and therefore cannot account for any size effects, such as effects of fiber diameter or fiber spacing, observed in real fiber composites.To illustrate the theoretical basis for expecting size effects, it is sufficient to consider pure bending in plane strain of a linear elastic plate with variable Young's modulus. In terms of rectangular Cartesian coordinates (x, y, z), suppose the middle surface of the plate lies in the plane $y = 0$, and the lateral surfaces are $y = \pm h$. Assuming that the plate undergoes a pure bending deformation, as in the elementary Euler–Bernoulli bending theory, the displacement in the x direction is

$$u = \frac{x\,y}{R}, \tag{5.1}$$

where R is the radius of curvature of the deformed plate. Further, suppose that the extension modulus E in the x direction depends on y, so that $E = E(y)$, with mean value E_0, where

$$E_0 = \frac{1}{2h}\int_{-h}^{h} E(y)\mathrm{d}y. \tag{5.2}$$

Then the stress component σ_{xx} is

$$\sigma_{xx} = E(y)\frac{\partial u}{\partial x} = E(y)\frac{y}{R}, \tag{5.3}$$

and the bending moment applied to a section $x = $ const., per unit length in the z direction, is

$$M_z = \int_{-h}^{h} \sigma_{xx} y\mathrm{d}y = \frac{1}{R}\int_{-h}^{h} E(y)y^2\mathrm{d}y, \tag{5.4}$$

and so the bending stiffness B is

$$B = M_z R = \int_{-h}^{h} E(y)y^2\mathrm{d}y. \tag{5.5}$$

On the other hand, if the extension modulus has the constant value E_0 the bending stiffness is

$$B_0 = \int_{-h}^{h} E_0 y^2\mathrm{d}y = \frac{2}{3}E_0 h^3, \tag{5.6}$$

Clearly, in general B and B_0 are not the same. To take a simple example for illustration, let

$$E(y) = E_0 - E_1 \cos\frac{\pi y}{d} \text{ where } d = h/N, \tag{5.7}$$

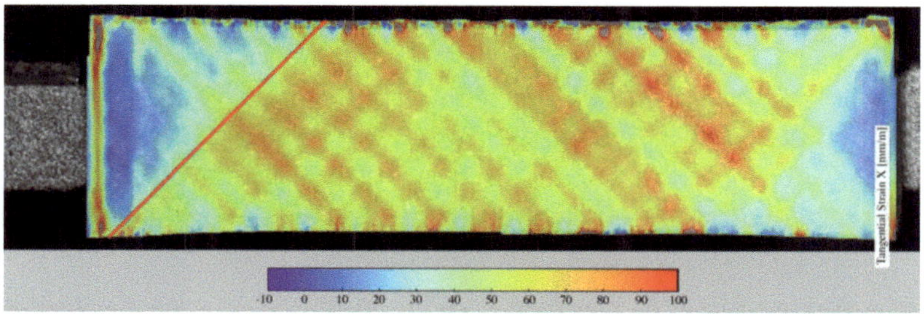

Figure 5.1: DIC measurement of a 25 [mm] specimen in 45° configuration, colors indicate local stretches in horizontal direction in [mm/m].

and N is an odd integer. Thus d is a measure of the scale of the inhomogeneity of the material. Then from (5.5)

$$B = B_0 + \frac{4d^2hE_1}{\pi^2} = B_0 + B_1\frac{d^2}{h^2}, \quad B_1 = \frac{6E_1}{E_0\pi^2}B_0, \tag{5.8}$$

and so the bending stiffness differs from B_0 by a term of order $(d/h)^2$. In the theory of classical fiber-reinforced materials, it is explicitly assumed that the fibers are infinitely thin and thus infinitely flexible, with the bending stiffness for a single fiber (represented by a mathematical curve in the theory) being zero. This clearly corresponds to the limit $d/h \to 0$ in (5.7). To relax this assumption while remaining within a continuum theory, it is necessary to introduce a length scale into the theory that effectively endows the fibers with bending stiffness.

The in-plane curvature resistance of fibers exhibits the same microstructural size effects. The digital image correlation (DIC) of a thermoplastic composite laminate, also known as an organic sheet, is shown in Figure 5.1.The additional penalty imposed by the fiber in-plane flexure is seen in the deformation patterns of the DIC measurement in comparison to the straight red line drawn in the figure. The chapter's proposed computational model for advanced mechanical modeling of fiber materials will be validated using the aforementioned prototypical composite material.

5.1.2 Kinematics

We consider a fiber-reinforced composite as a Kirchhoff-Love shell, as introduced in Section 2.3 with the therein presented configuration and enhanced kinematics. The derived standard deformation measures $a_{\alpha\beta}$ and $\kappa_{\alpha\beta}$ with respect to stretch and out-of-plane curvature and the additional measure $S^{\sigma}_{\alpha\beta}$ for in-plane curvature are first examined for material symmetry and objectivity. Furthermore, specific deformation measures for the embedded fibers are derived and discussed.

For the matrix material, we use the standard Kirchhoff-Love strain measure, namely the right Cauchy-Green tensor and the in-plane Jacobian determinant, as outlined in Section

2.3

$$\hat{\boldsymbol{C}} = \hat{\boldsymbol{F}}^{\mathrm{T}}\hat{\boldsymbol{F}} = g_{ij}\boldsymbol{G}^i \otimes \boldsymbol{G}^j \ , \tag{5.9}$$

and

$$\hat{J}_0(\theta^i) := \sqrt{\det(\hat{\boldsymbol{C}}(\theta^i))} = \sqrt{\frac{\det(g_{\alpha\beta}(\theta^\alpha,\theta^3))}{\det(G_{\alpha\beta}(\theta^\alpha,\theta^3))}}. \tag{5.10}$$

Frame invariance

The present formulation is based on the work of Steigmann [116], who used the Murdoch–Cohen [94] definition of material symmetry to derive the canonical forms of the strain-energy functions for woven fabrics with bending resistance. In the following, the invariance of the introduced measures $\{a_{\alpha\beta}, \kappa_{\alpha,\beta}, S^{\sigma}_{\alpha,\beta}\}$ under superposed rigid body motions will be briefly demonstrated. This includes translational as well as rotational invariance of the strain energy which in turn implies satisfaction for conservation of linear and angular momentum, see Hesch & Betsch [58] and Marsden & Ratiu [79] for further details.

We define rigid motions of the form

$$\boldsymbol{r}^+ := \boldsymbol{u} + \boldsymbol{Q}\boldsymbol{r} \ , \tag{5.11}$$

where $\boldsymbol{u} \in \mathbb{E}^3$ and $\boldsymbol{Q} \in SO(3)$ is a rotation tensor with the property $\boldsymbol{Q}^{\mathrm{T}}\boldsymbol{Q} = \boldsymbol{I}$. Then, the first and second partial derivative read

$$\boldsymbol{a}^+_\alpha = \boldsymbol{Q}\boldsymbol{r}_{,\alpha} \quad \text{and} \quad \boldsymbol{a}^+_{\alpha\beta} = \boldsymbol{Q}\boldsymbol{r}_{,\alpha\beta} \ . \tag{5.12}$$

From the covariant metric coefficients we obtain

$$a^+_{\alpha\beta} = \boldsymbol{a}^+_\alpha \cdot \boldsymbol{a}^+_\beta = (\boldsymbol{Q}\boldsymbol{r}_{,\alpha}) \cdot (\boldsymbol{Q}\boldsymbol{r}_{,\beta}) = \boldsymbol{r}_{,\alpha} \cdot \underbrace{\boldsymbol{Q}^{\mathrm{T}}\boldsymbol{Q}}_{=\boldsymbol{I}}\, \boldsymbol{r}_{,\beta} = a_{\alpha\beta} \tag{5.13}$$

which clearly holds also for the contravariant metric coefficients, i.e. $(a^{\alpha\beta})^+ = a^{\alpha\beta}$. With regards to this finding, we can demonstrate

$$S^+_{\alpha\beta\lambda} = \Gamma^+_{\alpha\beta\lambda} - \bar{\Gamma}^\lambda_{\alpha\beta} a^+_{\alpha\lambda} = (\boldsymbol{Q}\boldsymbol{r}_{,\lambda}) \cdot (\boldsymbol{Q}\boldsymbol{r}_{,\alpha\beta}) - \bar{\Gamma}^\lambda_{\alpha\beta} a_{\alpha\lambda} = \boldsymbol{r}_{,\lambda} \cdot \underbrace{\boldsymbol{Q}^{\mathrm{T}}\boldsymbol{Q}}_{=\boldsymbol{I}}\, \boldsymbol{r}_{,\alpha\beta} - \bar{\Gamma}^\lambda_{\alpha\beta} a_{\alpha\lambda} = S_{\alpha\beta\lambda} \tag{5.14}$$

as well as

$$(S^\sigma_{\alpha\beta})^+ = S^+_{\alpha\beta\lambda}(a^{\sigma\lambda})^+ = S_{\alpha\beta\lambda}a^{\sigma\lambda} = S^\sigma_{\alpha\beta} \ , \tag{5.15}$$

i.e. the frame invariance of the co- and contravariant in-plane curvature tensor is confirmed. Eventually, we obtain for the out-of-plane curvature tensor

$$\begin{aligned}
\kappa^+_{\alpha\beta} &= B_{\alpha\beta} - b^+_{\alpha\beta} \\
&= B_{\alpha\beta} - \boldsymbol{Q}\boldsymbol{r}_{,\alpha\beta} \cdot \frac{\boldsymbol{Q}\boldsymbol{a}_1 \times \boldsymbol{Q}\boldsymbol{a}_2}{\|\boldsymbol{Q}\boldsymbol{a}_1 \times \boldsymbol{Q}\boldsymbol{a}_2\|} \\
&= B_{\alpha\beta} - \frac{\det(\boldsymbol{Q}\boldsymbol{r}_{,\alpha\beta}, \boldsymbol{Q}\boldsymbol{a}_1, \boldsymbol{Q}\boldsymbol{a}_2)}{\sqrt{\det(a^+_{\alpha\beta})}} \\
&= B_{\alpha\beta} - \underbrace{\det(\boldsymbol{Q})}_{=1} \frac{\det(\boldsymbol{r}_{,\alpha\beta}, \boldsymbol{a}_1, \boldsymbol{a}_2)}{\sqrt{\det(a_{\alpha\beta})}} \\
&= \kappa_{\alpha\beta} \, .
\end{aligned} \tag{5.16}$$

confirming again the frame invariance.

Fiber deformation measures

Following Steigmann and dell'Isola [117], we now derive additional deformation measures for the embedded fibers which all can be considered as functions of $a_{\alpha\beta}$, $S^{\sigma}_{\alpha\beta}$, $\kappa_{\alpha\beta}$ and θ^{α}. Thus, we focus on the mathematical modeling of fibers as curves with appropriate kinematical and constitutive structure to capture the mechanical properties of the reinforcement.

The two fiber threads interlaced at right angles to each other, the so-called warp and weft threads of the woven fabric, are characterized by their constant orthogonal vector fields in the reference configuration $\boldsymbol{M}$ and $\boldsymbol{L}$. Described in the reference mid-surface Ω, and not necessarily with unit length, they can be written as

$$\boldsymbol{L} = L^{\alpha}\boldsymbol{A}_{\alpha} \quad \text{and} \quad \boldsymbol{M} = M^{\alpha}\boldsymbol{A}_{\alpha} \, . \tag{5.17}$$

On the deformed mid-surface ω they are denoted by $\boldsymbol{l}$ and $\boldsymbol{m}$ and can be represented as

$$\boldsymbol{l} = l^{\alpha}\boldsymbol{a}_{\alpha} \quad \text{and} \quad \boldsymbol{m} = m^{\alpha}\boldsymbol{a}_{\alpha} \, . \tag{5.18}$$

Since the fibers are convected as material curves, the tangent vectors are related by

$$\lambda_1\boldsymbol{l} = \boldsymbol{F}\frac{\boldsymbol{L}}{\|\boldsymbol{L}\|} \quad \text{and} \quad \lambda_2\boldsymbol{l} = \boldsymbol{F}\frac{\boldsymbol{M}}{\|\boldsymbol{M}\|} \tag{5.19}$$

where we made use of the fiber stretches

$$\lambda_1 := \left\| \frac{\boldsymbol{F}\boldsymbol{L}}{\|\boldsymbol{L}\|} \right\| = \frac{\sqrt{a_{\alpha\beta}L^{\alpha}L^{\beta}}}{\sqrt{A_{\mu\nu}L^{\mu}L^{\nu}}} \quad \text{and} \quad \lambda_2 := \left\| \frac{\boldsymbol{F}\boldsymbol{M}}{\|\boldsymbol{M}\|} \right\| = \frac{\sqrt{a_{\alpha\beta}M^{\alpha}M^{\beta}}}{\sqrt{A_{\mu\nu}M^{\mu}M^{\nu}}} \, . \tag{5.20}$$

The corresponding contravariant components are connected by

$$L^\alpha = \lambda_1 \|\boldsymbol{L}\| l^\alpha = \sqrt{a_{\mu\nu} L^\mu L^\nu} l^\alpha \quad \text{and} \quad M^\alpha = \lambda_2 \|\boldsymbol{M}\| M^\alpha = \sqrt{a_{\mu\nu} M^\mu M^\nu} m^\alpha \,. \tag{5.21}$$

Furthermore, the change of angle between the fibers can be formulated as

$$\varphi := \operatorname{acos}\left(\frac{A_{\alpha\beta} L^\alpha M^\beta}{\sqrt{A_{\mu\nu} L^\mu L^\nu A_{\sigma\rho} M^\sigma M^\rho}}\right) - \operatorname{acos}\left(\frac{a_{\alpha\beta} L^\alpha M^\beta}{\sqrt{a_{\mu\nu} L^\mu L^\nu a_{\sigma\rho} M^\sigma M^\rho}}\right). \tag{5.22}$$

Hereby, the first term is $\pi/2$ for initially orthogonal fibers.

To study the change of the tangent vectors $\boldsymbol{l}$ and $\boldsymbol{m}$ within the deformed mid-surface, we introduce the auxiliary unit tangent vectors

$$\boldsymbol{p} = \boldsymbol{n} \times \boldsymbol{l} \,, \quad \boldsymbol{q} = \boldsymbol{n} \times \boldsymbol{m} \,. \tag{5.23}$$

Moreover, let $\tilde{\boldsymbol{l}} \colon \mathcal{B} \to \mathbb{E}^3$ and $\tilde{\boldsymbol{m}} \colon \mathcal{B} \to \mathbb{E}^3$ be such that

$$\boldsymbol{l}(\theta^\alpha) = \tilde{\boldsymbol{l}}(\boldsymbol{r}(\boldsymbol{\theta}^\alpha)) \quad \text{and} \quad \boldsymbol{m}(\theta^\alpha) = \tilde{\boldsymbol{m}}(\boldsymbol{r}(\boldsymbol{\theta}^\alpha)) \,. \tag{5.24}$$

Then the gradients

$$\nabla\tilde{\boldsymbol{l}}(\boldsymbol{r}) = \boldsymbol{l}_{,\alpha} \otimes \boldsymbol{a}^\alpha \quad \text{and} \quad \nabla\tilde{\boldsymbol{m}}(\boldsymbol{r}) = \boldsymbol{m}_{,\alpha} \otimes \boldsymbol{a}^\alpha \tag{5.25}$$

are easily obtained by inserting $\mathrm{d}\theta^\alpha = \mathrm{d}\boldsymbol{r} \cdot \boldsymbol{a}^\alpha$ into the expressions $\mathrm{d}\boldsymbol{l} = \boldsymbol{l}_{,\alpha}\,\mathrm{d}\theta^\alpha = \boldsymbol{l}_{,\alpha}(\,\mathrm{d}\boldsymbol{r} \cdot \boldsymbol{a}^\alpha) = (\boldsymbol{l}_{,\alpha} \otimes \boldsymbol{a}^\alpha)\,\mathrm{d}\boldsymbol{r}$ and $\mathrm{d}\boldsymbol{m} = \boldsymbol{m}_{,\alpha}\,\mathrm{d}\theta^\alpha = (\boldsymbol{m}_{,\alpha} \otimes \boldsymbol{a}^\alpha)\,\mathrm{d}\boldsymbol{r}$, respectively. Since $\|\boldsymbol{l}\| = \|\boldsymbol{m}\| = 1$, we have $\boldsymbol{l} \cdot \boldsymbol{l}_{,\alpha} = \boldsymbol{m} \cdot \boldsymbol{m}_{,\alpha} = 0$ and the partial derivatives of the unit vectors $\boldsymbol{l}$ and $\boldsymbol{m}$ can be expressed in the right-handed orthonormal bases $\{\boldsymbol{l}, \boldsymbol{p}, \boldsymbol{n}\}$ and $\{\boldsymbol{m}, \boldsymbol{q}, \boldsymbol{n}\}$ as

$$\boldsymbol{l}_{,\alpha} = (\boldsymbol{l}_{,\alpha} \cdot \boldsymbol{p})\boldsymbol{p} + (\boldsymbol{l}_{,\alpha} \cdot \boldsymbol{n})\boldsymbol{n} \quad \text{and} \quad \boldsymbol{m}_{,\alpha} = (\boldsymbol{m}_{,\alpha} \cdot \boldsymbol{q})\boldsymbol{q} + (\boldsymbol{m}_{,\alpha} \cdot \boldsymbol{n})\boldsymbol{n}. \tag{5.26}$$

Using (5.18) together with the Gauss and Weingarten equation (2.34), the normal components in (5.26) can be expressed using the covariant components of the mid-surface curvature $b_{\alpha\beta}$ as

$$\boldsymbol{l}_{,\alpha} \cdot \boldsymbol{n} = (l^\beta \boldsymbol{a}_\beta)_{,\alpha} \cdot \boldsymbol{n} = l^\beta b_{\beta\alpha} \quad \text{and} \quad \boldsymbol{m}_{,\alpha} \cdot \boldsymbol{n} = (m^\beta \boldsymbol{a}_\beta)_{,\alpha} \cdot \boldsymbol{n} = m^\beta b_{\beta\alpha} \,. \tag{5.27}$$

The change of the unit tangent vectors in tangential direction can therefore be represented by

$$\nabla\tilde{\boldsymbol{l}}(\boldsymbol{r})\boldsymbol{l} = l^\alpha \boldsymbol{l}_{,\alpha} = \eta_l \boldsymbol{p} + \kappa_l \boldsymbol{n} \quad \text{and} \quad \nabla\tilde{\boldsymbol{m}}(\boldsymbol{r})\boldsymbol{m} = m^\alpha \boldsymbol{m}_{,\alpha} = \eta_m \boldsymbol{q} + \kappa_m \boldsymbol{n} \tag{5.28}$$

with the normal fiber curvatures

$$\kappa_l := b_{\alpha\beta} l^\alpha l^\beta = \frac{b_{\alpha\beta} L^\alpha L^\beta}{a_{\mu\nu} L^\mu L^\nu} \quad \text{and} \quad \kappa_m := b_{\alpha\beta} m^\alpha m^\beta = \frac{b_{\alpha\beta} M^\alpha M^\beta}{a_{\mu\nu} M^\mu M^\nu} \tag{5.29}$$

and the geodesic curvatures

$$\eta_l := l^\alpha \boldsymbol{l}_{,\alpha} \cdot \boldsymbol{p} \quad \text{and} \quad \eta_m := m^\alpha \boldsymbol{m}_{,\alpha} \cdot \boldsymbol{q} \,. \tag{5.30}$$

The change of the unit tangent vectors of one fiber family in direction of the other fiber family leads to

$$\nabla\tilde{\boldsymbol{l}}(\boldsymbol{r})\boldsymbol{m} = m^\alpha \boldsymbol{l}_{,\alpha} = \phi_l \boldsymbol{p} + \tau \boldsymbol{n} \quad \text{and} \quad \nabla\tilde{\boldsymbol{m}}(\boldsymbol{r})\boldsymbol{m} = m^\alpha \boldsymbol{m}_{,\alpha} = \phi_m \boldsymbol{q} + \tau \boldsymbol{n} \,, \tag{5.31}$$

where we recognize the torsion and the respective Tchebychev curvatures, [118],

$$\tau := b_{\alpha\beta} l^\alpha m^\beta = \frac{b_{\alpha\beta} L^\alpha M^\beta}{\sqrt{a_{\mu\nu} L^\mu L^\nu a_{\sigma\rho} M^\sigma M^\rho}} \,, \quad \phi_l := m^\alpha (\boldsymbol{l}_{,\alpha} \cdot \boldsymbol{p}) \,, \quad \phi_m := l^\alpha (\boldsymbol{m}_{,\alpha} \cdot \boldsymbol{q}) \,. \tag{5.32}$$

Similarly to (5.29) and (5.32), we can introduce the normal fiber curvatures κ_L, κ_M and the torsion $\bar{\tau}$ on the reference mid-surface as

$$\kappa_L := \frac{B_{\alpha\beta} L^\alpha L^\beta}{A_{\mu\nu} L^\mu L^\nu} \,, \quad \kappa_M := \frac{B_{\alpha\beta} M^\alpha M^\beta}{A_{\mu\nu} M^\mu M^\nu} \quad \text{and} \quad \bar{\tau} := \frac{B_{\alpha\beta} L^\alpha M^\beta}{\sqrt{A_{\mu\nu} L^\mu L^\nu A_{\sigma\rho} M^\sigma M^\rho}} \,. \tag{5.33}$$

Because these definitions rely on unit tangent vectors, the normalization by $||\boldsymbol{L}||$ and $||\boldsymbol{M}||$ are involved in here.

As strain measures that captures the change in the normal curvature of the fibers, i.e. by out-of-plane bending, we choose

$$K_1 := -\frac{\kappa_{\alpha\beta} L^\alpha L^\beta}{A_{\mu\nu} L^\mu L^\nu} = \lambda_1^2 \kappa_l - \kappa_L \quad \text{and} \quad K_2 := -\frac{\kappa_{\alpha\beta} M^\alpha M^\beta}{A_{\mu\nu} M^\mu M^\nu} = \lambda_2^2 \kappa_m - \kappa_M \,. \tag{5.34}$$

The change in torsion by twisting the fibers is given by

$$K_3 := -\frac{\kappa_{\alpha\beta} L^\alpha M^\beta}{\sqrt{A_{\mu\nu} L^\mu L^\nu A_{\sigma\rho} M^\sigma M^\rho}} = \lambda_1 \lambda_2 \tau - \bar{\tau} \,. \tag{5.35}$$

Finally, to describe the deformation of the fibers within the tangent planes of the mid-surfaces, like in-plane-bending, we introduce as strain measures

$$\boldsymbol{g}_L := \frac{L^\alpha L^\beta S_{\alpha\beta\sigma}}{A_{\mu\nu} L^\mu L^\nu} \boldsymbol{a}^\sigma \,, \quad \boldsymbol{g}_M := \frac{M^\alpha M^\beta S_{\alpha\beta\sigma}}{A_{\mu\nu} M^\mu M^\nu} \boldsymbol{a}^\sigma \quad \text{and} \quad \boldsymbol{\Gamma} := \frac{L^\alpha M^\beta S_{\alpha\beta\sigma}}{\sqrt{A_{\mu\nu} L^\mu L^\nu A_{\sigma\rho} M^\sigma M^\rho}} \boldsymbol{a}^\sigma \,. \tag{5.36}$$

For initially straight fibers, the deformation measures (5.36) can be represented as

$$\begin{aligned} \boldsymbol{g}_L &= \lambda_1 \eta_l \boldsymbol{p} + (\boldsymbol{L} \cdot \nabla\lambda_1) \boldsymbol{l} \,, \quad \boldsymbol{g}_M = \lambda_2 \eta_m \boldsymbol{q} + (\boldsymbol{M} \cdot \nabla\lambda_2) \boldsymbol{m} \\ \boldsymbol{\Gamma} &= (\boldsymbol{L} \cdot \nabla\lambda_2) \boldsymbol{m} + \lambda_1 \lambda_2 \phi_m \boldsymbol{q} = (\boldsymbol{M} \cdot \nabla\lambda_1) \boldsymbol{l} + \lambda_1 \lambda_2 \phi_l \boldsymbol{p} \,, \end{aligned} \tag{5.37}$$

with the stretch gradients as $\nabla\lambda_1 = \lambda_{1,\alpha} \boldsymbol{A}^\alpha$ and $\nabla\lambda_2 = \lambda_{2,\alpha} \boldsymbol{A}^\alpha$ (see Equation (58) and (59) in [117]). We refer to Section 2.4 of [117], where also the general precurved situation is analyzed. However, the restriction to straight fibers lies only in the representation (5.37), which should help to get more insight into strain measures (5.36). In fact, these strain measures obtained by the contraction of the fiber tangents of the reference mid-surface with the third-order tensor $\boldsymbol{a}_\sigma \otimes \boldsymbol{S}^\sigma$ represent a combination of the gradients of the fiber stretch and in-plane-deformation of the fibers described by geodesic and Tchebychev curvatures.

5.1.3 Variational formulation

Energetic response

For the matrix material we assume the strain energy function per unit reference volume $\Psi^{\text{iso}}(\hat{C}_{ij}, \theta^i)$, for the fibers per unit reference area $\Psi^{\text{fib}}(a_{\alpha\beta}, S_{\alpha\beta\sigma}, \kappa_{\alpha\beta}, \theta^\alpha)$. Moreover, we assume that the embedded fibers have a height of h_{f} such that the space occupied by matrix material in the reference configuration is defined by $\{\boldsymbol{X}(\theta^i)|\theta^3 \in \varpi_{\text{iso}}\}$, where $\varpi_{\text{iso}} = \left[-\frac{h}{2}, \frac{h}{2}\right] \setminus \left(-\frac{h_{\text{f}}}{2}, \frac{h_{\text{f}}}{2}\right)$. For the overall composite we obtain for the strain energy defined per unit reference area of the surface

$$\Psi(a_{\alpha\beta}, S_{\alpha\beta\sigma}, \kappa_{\alpha\beta}, \theta^\alpha) = \int\limits_{\varpi_{\text{iso}}} \Psi^{\text{iso}}(\hat{C}_{ij}(a_{\alpha\beta}, \kappa_{\alpha\beta}, \theta^3), \theta^i)\, \mathrm{d}\theta^3 + \Psi^{\text{fib}}(a_{\alpha\beta}, S_{\alpha\beta\sigma}, \kappa_{\alpha\beta}, \theta^\alpha)\ . \quad (5.38)$$

Assuming a plane stress distribution in the three-dimensional continuum together with a decoupling of in-plane and thickness deformations, we use this specific incompressible neo-Hookean strain energy function for the matrix material

$$\Psi^{\text{iso}} = \frac{1}{2}\mu\left(\operatorname{tr}(\hat{C}_{\alpha\beta}\boldsymbol{G}^\alpha \otimes \boldsymbol{G}^\beta) + \left(\frac{1}{\hat{J}_0}\right)^2 - 3\right) = \frac{1}{2}\mu\left(\hat{C}_{\alpha\beta}G^{\alpha\beta} + \frac{\det(G_{\mu\nu})}{\det(\hat{C}_{\alpha\beta})} - 3\right), \quad (5.39)$$

with the shear modulus μ. Note that this is a correction on a constitutive level for the too restrictive kinematical ansatz 2.31, which does not allow for thickness deformation. In the numerical examples, we will also consider a compressible neo-Hookean matrix material, given by

$$\Psi^{\text{iso}} = \frac{1}{2}\mu(J^{-2/3}\operatorname{tr}(\hat{\boldsymbol{C}}) - 3) + \frac{1}{4}\kappa(J^2 - 1 - 2\ln(J)). \quad (5.40)$$

Since the plane stress condition cannot be inserted analytically here, we have to condense the corresponding conditions numerically. For details we refer to Kiendl et al. [70].

Assuming rectangular cross-sections for the embedded fibers, we adopt the following strain energy function

$$\Psi^{\text{fib}} = \sum_{\alpha=1}^{2} \frac{a_\alpha}{2}(\lambda_\alpha - 1)^2 + \frac{a_3}{2}\tan(\varphi)^2 + \sum_{i=1}^{3} \frac{k_i}{2}(K_i)^2 + \frac{1}{2}(g_1(\boldsymbol{g}_L \cdot \boldsymbol{g}_L) + g_2(\boldsymbol{g}_M \cdot \boldsymbol{g}_M) + g_3(\boldsymbol{\Gamma} \cdot \boldsymbol{\Gamma})) \quad (5.41)$$

where the constant material parameters a_i are associated to the stretch and the change of angle between the fibers. In particular, $a_\alpha = E_\alpha h_{\text{f}}$ correlates to Young's modulus E_α of the particular fiber, whereas $a_3 = G h_{\text{f}}$ represents the shear stiffness of a shell continuum. Note that the fiber formulation does not take any Poisson effect into account, i.e. the Poisson's ratio $\nu = 0$ and thus, $G = \frac{1}{2}E$ in a classical continuum. In the present formulation, G represents the stiffness against twist between the fibers and is in general independent of the fiber stiffness. The out-of-plane bending stiffness k_α depends on the particular microstructure of the embedded fibers, i.e. on the second moment of area, such that $k_\alpha = E_\alpha I_\alpha / b_\alpha$. Here, b_α denotes a representative in-plane length scale of the fiber, such that we obtain a bending stiffness of $k_\alpha = E_\alpha h_\mathrm{f}^3/12$ for a continuously distributed fiber. Analogously, we obtain for the in-plane bending stiffness $g_\alpha = E_\alpha b_\alpha^2 h_\mathrm{f}/12$. Note that other microstructures are possible, although characteristics like fiber height and width have to be discussed and validated for the particular geometry. Eventually, the terms associated with k_3 and g_3 control the fiber torsion as well as the in-plane deformation captured by Tchebychev curvatures and strain-gradients, see (5.37). Assuming that the same fibers are used in both directions, the same material parameters can be applied for both fibers, i.e. $a_1 = a_2$, $k_1 = k_2$ and $g_1 = g_2$. Moreover, we note that the used strain measures of the fiber components associated with curvature terms account always for combined effects of the gradients of the fiber stretch and bending. Thus, the identification of the material parameters in terms of classical beam theory is limited, as we will show in Section 3.5.

Principle of virtual work

With the strain energy function presented, we can define the internal energy as

$$W^\mathrm{int} = \int_\Omega \Psi \, \mathrm{d}A. \tag{5.42}$$

For the finite element analysis, we discretize the virtual work expression as

$$\delta W^\mathrm{int} = \int_\Omega \left[\int_{\varpi_\mathrm{iso}} \delta\Psi^\mathrm{iso}(\hat{C}_{ij}(a_{\alpha\beta}, \kappa_{\alpha\beta}, \theta^3), \theta^i) \, \mathrm{d}\theta^3 + \delta\Psi^\mathrm{fib}(a_{\alpha\beta}, S_{\alpha\beta\sigma}, \kappa_{\alpha\beta}, \theta^\alpha) \right] \mathrm{d}A\,. \tag{5.43}$$

Introducing the stress resultants for the matrix and fiber contributions directly related to the mid-surface, respectively as

$$n_\mathrm{fib}^{\alpha\beta} := \frac{\partial \Psi^\mathrm{fib}}{\partial a_{\alpha\beta}}\,, \quad m_\mathrm{fib}^{\alpha\beta} := \frac{\partial \Psi^\mathrm{fib}}{\partial \kappa_{\alpha\beta}} \quad \text{and} \quad m_\mathrm{fib}^{\alpha\beta\sigma} := \frac{\partial \Psi^\mathrm{fib}}{\partial S_{\alpha\beta\sigma}} \tag{5.44}$$

and

$$n_\mathrm{iso}^{\alpha\beta} := \int_{\varpi_\mathrm{iso}} \frac{\partial \Psi^\mathrm{iso}}{\partial \hat{C}_{\alpha\beta}} \mathrm{d}\theta^3 \quad \text{and} \quad m_\mathrm{iso}^{\alpha\beta} := \int_{\varpi_\mathrm{iso}} 2\theta^3 \frac{\partial \Psi^\mathrm{iso}}{\partial \hat{C}_{\alpha\beta}} \mathrm{d}\theta^3, \tag{5.45}$$

along with $\delta\hat{C}_{ij} = \delta_i^\alpha \delta_j^\beta (\delta a_{\alpha\beta} + 2\theta^3 \delta\kappa_{\alpha\beta})$ yields

$$\delta W^\mathrm{int} = \int_\Omega \left[(n_\mathrm{iso}^{\alpha\beta} + n_\mathrm{fib}^{\alpha\beta})\delta a_{\alpha\beta} + (m_\mathrm{iso}^{\alpha\beta} + m_\mathrm{fib}^{\alpha\beta})\delta\kappa_{\alpha\beta} + m_\mathrm{fib}^{\alpha\beta\sigma}\delta S_{\alpha\beta\sigma} \right] \mathrm{d}A\,. \tag{5.46}$$

Applying the strain energy function of the matrix material given in (5.39) and using $[\hat{C}^{\alpha\beta}] = [\hat{C}_{\alpha\beta}]^{-1}$ together with

$$\frac{\partial}{\partial \hat{C}_{\mu\nu}} \left(\frac{1}{\det(\hat{C}_{\alpha\beta})} \right) = -\frac{1}{\det(\hat{C}_{\alpha\beta})^2} \det(\hat{C}_{\lambda\sigma}) \hat{C}^{\mu\nu} = -\frac{\hat{C}^{\mu\nu}}{\det(\hat{C}_{\alpha\beta})} \tag{5.47}$$

we get

$$\frac{\partial \Psi^{\text{iso}}}{\partial \hat{C}_{\mu\nu}} = \frac{1}{2}\mu\left(G^{\mu\nu} - \frac{\det(G_{\mu\nu})}{\det(\hat{C}_{\alpha\beta})}\hat{C}^{\mu\nu}\right) \tag{5.48}$$

that has to be integrated in thickness direction to obtain the stress resultants.

The corresponding stress resultants of the first term $(5.44)_1$ reads

$$\begin{aligned} n^{\alpha\beta}_{\text{fib}} =& \frac{\partial \Psi^{\text{fib}}}{\partial \lambda_1}\frac{\partial \lambda_1}{\partial a_{\alpha\beta}} + \frac{\partial \Psi^{\text{fib}}}{\partial \lambda_2}\frac{\partial \lambda_2}{\partial a_{\alpha\beta}} + \frac{\partial \Psi^{\text{fib}}}{\partial \varphi}\frac{\partial \varphi}{\partial a_{\alpha\beta}} + \frac{\partial \Psi^{\text{fib}}}{\partial C^{\sigma\mu}}\frac{\partial C^{\sigma\mu}}{\partial a_{\alpha\beta}} \\ =& a_1(\lambda_1 - 1)\frac{\partial \lambda_1}{\partial a_{\alpha\beta}} + a_2(\lambda_2 - 1)\frac{\partial \lambda_2}{\partial a_{\alpha\beta}} + 2a_3\frac{\tan(\varphi)}{\cos^2(\varphi)}\frac{\partial \varphi}{\partial a_{\alpha\beta}} \\ &+ \left(g_1\frac{(L^\gamma L^\lambda S_{\gamma\lambda\sigma})(L^\varsigma L^\iota S_{\varsigma\iota\mu})}{(A_{\mu\nu}L^\mu L^\nu)^2} + g_2\frac{(M^\gamma M^\lambda S_{\gamma\lambda\sigma})(M^\varsigma M^\iota S_{\varsigma\iota\mu})}{(A_{\mu\nu}M^\mu M^\nu)^2}\right. \\ &+ \left. g_3\frac{(L^\gamma M^\lambda S_{\gamma\lambda\sigma})(L^\varsigma M^\iota S_{\varsigma\iota\mu})}{A_{\mu\nu}L^\mu L^\nu A_{\sigma\rho}M^\sigma M^\rho}\right)\frac{\partial C^{\sigma\mu}}{\partial a_{\alpha\beta}}\,, \end{aligned} \tag{5.49}$$

where we have made use of (5.41) as strain energy function of the fiber material and

$$\begin{aligned} \boldsymbol{g}_L \cdot \boldsymbol{g}_L =& \frac{(L^\alpha L^\beta S_{\alpha\beta\sigma})(L^\gamma L^\lambda S_{\gamma\lambda\mu})C^{\sigma\mu}}{(A_{\mu\nu}L^\mu L^\nu)^2}\,, \\ \boldsymbol{g}_M \cdot \boldsymbol{g}_M =& \frac{(M^\alpha M^\beta S_{\alpha\beta\sigma})(M^\gamma M^\lambda S_{\gamma\lambda\mu})C^{\sigma\mu}}{(A_{\mu\nu}M^\mu M^\nu)^2}\,, \\ \boldsymbol{\Gamma} \cdot \boldsymbol{\Gamma} =& \frac{(L^\alpha M^\beta S_{\alpha\beta\sigma})(L^\gamma M^\lambda S_{\gamma\lambda\mu})C^{\sigma\mu}}{A_{\mu\nu}L^\mu L^\nu A_{\sigma\rho}M^\sigma M^\rho}\,. \end{aligned} \tag{5.50}$$

Moreover, the relations

$$\frac{\partial \lambda_1}{\partial a_{\alpha\beta}} = \frac{1}{2}\frac{L^\alpha L^\beta}{\sqrt{a_{\sigma\lambda}L^\sigma L^\lambda A_{\mu\nu}L^\mu L^\nu}}\,, \quad \frac{\partial \lambda_2}{\partial a_{\alpha\beta}} = \frac{1}{2}\frac{M^\alpha M^\beta}{\sqrt{a_{\sigma\lambda}M^\sigma M^\lambda A_{\mu\nu}M^\mu M^\nu}} \tag{5.51}$$

and

$$\begin{aligned} \frac{\partial \varphi}{a_{\alpha\beta}} =& \frac{1}{\sqrt{1 - \frac{(a_{\alpha\beta}L^\alpha M^\beta)^2}{a_{\mu\nu}L^\mu L^\nu a_{\sigma\rho}M^\sigma M^\rho}}}\left(\frac{L^\alpha M^\beta}{\sqrt{a_{\mu\nu}L^\mu L^\nu a_{\sigma\rho}M^\sigma M^\rho}}\right. \\ &- \left.\frac{a_{\alpha\beta}L^\alpha M^\beta}{2(a_{\mu\nu}L^\mu L^\nu a_{\sigma\rho}M^\sigma M^\rho)^{3/2}}(a_{\mu\nu}L^\mu L^\nu M^\sigma M^\rho + L^\mu L^\nu a_{\sigma\rho}M^\sigma M^\rho)\right) \end{aligned} \tag{5.52}$$

as well as the partial derivative of the contravariant metric

$$\frac{\partial C^{\sigma\mu}}{\partial a_{\alpha\beta}} = \frac{\partial C^{\sigma\mu}}{\partial C_{\alpha\beta}} = -C^{\sigma\alpha}C^{\beta\mu} \tag{5.53}$$

have been used. For the second term $(5.44)_2$, the stress resultants of the out-of-plane bending moments reads

$$m^{\alpha\beta}_{\text{fib}} = -k_1K_1\frac{L^\alpha L^\beta}{A_{\mu\nu}L^\mu L^\nu} - k_2K_2\frac{M^\alpha M^\beta}{A_{\mu\nu}M^\mu M^\nu} - k_3K_3\frac{L^\alpha M^\beta}{\sqrt{A_{\mu\nu}L^\mu L^\nu A_{\sigma\rho}M^\sigma M^\rho}}\,, \tag{5.54}$$

whereas the third term $(5.44)_3$ yields

$$\begin{aligned} m_{\text{fib}}^{\alpha\beta\sigma} =& g_1 \frac{L^\alpha L^\beta (L^\gamma L^\lambda S_{\gamma\lambda\mu})}{(A_{\mu\nu} L^\mu L^\nu)^2} C^{\sigma\mu} + g_2 \frac{M^\alpha M^\beta (M^\gamma M^\lambda S_{\gamma\lambda\mu})}{(A_{\mu\nu} M^\mu M^\nu)^2} C^{\sigma\mu} + \\ & g_3 \frac{L^\alpha M^\beta (L^\gamma M^\lambda S_{\gamma\lambda\mu})}{A_{\mu\nu} L^\mu L^\nu A_{\sigma\rho} M^\sigma M^\rho} C^{\sigma\mu} \, . \end{aligned} \tag{5.55}$$

The variations of $a_{\alpha\beta}$, $\kappa_{\alpha\beta}$ and $S_{\alpha\beta\sigma}$ can be expressed in terms of $\delta\boldsymbol{r}(\theta^\alpha) = \hat{\boldsymbol{r}}_{,\varepsilon}(\theta^\alpha, \varepsilon_0)$, which is the variational derivative of $\boldsymbol{r}$* induced by the one-parameter family $\hat{\boldsymbol{r}}(\theta^\alpha, \varepsilon)$ for which $\hat{\boldsymbol{r}}(\theta^\alpha, \varepsilon_0) = \boldsymbol{r}(\theta^\alpha)$ holds. The variation of the covariant base vectors as well as the partial derivatives thereof are $\delta\boldsymbol{a}_\alpha = \delta\boldsymbol{r}_{,\alpha}$ and $\delta\boldsymbol{a}_{\alpha,\beta} = \delta\boldsymbol{r}_{,\alpha\beta}$, respectively. The variation of the first fundamental form (2.33) can be written as

$$\delta a_{\alpha\beta} = \boldsymbol{a}_\alpha \cdot \delta\boldsymbol{a}_\beta + \boldsymbol{a}_\beta \cdot \delta\boldsymbol{a}_\alpha \, . \tag{5.56}$$

The variation of the out-of-plane curvature changes is due to (2.51) and (2.35) given by

$$\delta\kappa_{\alpha\beta} = -(\delta\boldsymbol{a}_{\alpha,\beta} \cdot \boldsymbol{n} + \boldsymbol{a}_{\alpha,\beta} \cdot \delta\boldsymbol{n}) \, , \tag{5.57}$$

where the variation of the unit normal vector reads

$$\delta\boldsymbol{n} = \frac{(\boldsymbol{I} - \boldsymbol{n} \otimes \boldsymbol{n})(\boldsymbol{a}_1 \times \delta\boldsymbol{a}_2 + \delta\boldsymbol{a}_1 \times \boldsymbol{a}_2)}{\|\boldsymbol{a}_1 \times \boldsymbol{a}_2\|} \tag{5.58}$$

using the identity map $\boldsymbol{I}$. Eventually, the variation of the in-plane curvature (2.48) gives rise to

$$\delta S_{\alpha\beta\sigma} = \delta\boldsymbol{a}_{\alpha,\beta} \cdot \boldsymbol{a}_\sigma + \boldsymbol{a}_{\alpha,\beta} \cdot \delta\boldsymbol{a}_\sigma - \bar{\Gamma}^\lambda_{\alpha\beta}(\boldsymbol{a}_\lambda \cdot \delta\boldsymbol{a}_\sigma + \boldsymbol{a}_\sigma \cdot \delta\boldsymbol{a}_\lambda) \, . \tag{5.59}$$

Strong and weak form

To identify and collect the resulting bending moments and normal stress contributions, we reorganize the virtual work expression. First, we reorganize the variation of the curvature tensor as follows

$$\begin{aligned} \delta b_{\alpha\beta} &= \boldsymbol{n} \cdot \delta\boldsymbol{a}_{\alpha,\beta} + \boldsymbol{a}_{\alpha,\beta} \cdot \delta\boldsymbol{n} \, , \\ &= \boldsymbol{n} \cdot \left[\delta\boldsymbol{a}_{\alpha,\beta} - \Gamma^\sigma_{\alpha\beta} \delta\boldsymbol{a}_\sigma\right] \, , \end{aligned} \tag{5.60}$$

where we make use of $\boldsymbol{a}_\sigma \cdot \delta\boldsymbol{n} = -\boldsymbol{n} \cdot \delta\boldsymbol{a}_\sigma$ and $\boldsymbol{n} \cdot \delta\boldsymbol{n} = 0$ along with the Gauss and Weingarten relation $(2.34)_1$. Taking the second covariant derivative introduced in (2.49) into account yields

$$\delta b_{\alpha\beta} = \boldsymbol{n} \cdot \left[\delta\boldsymbol{r}_{|\alpha\beta} - S^\sigma_{\alpha\beta} \delta\boldsymbol{a}_\sigma\right] \, . \tag{5.61}$$

Next, the covariant term $S_{\alpha\beta\sigma} = (\Gamma^\lambda_{\alpha\beta} - \bar{\Gamma}^\lambda_{\alpha\beta}) a_{\lambda\sigma}$ gives rise to

$$\delta S_{\alpha\beta\sigma} = \delta\Gamma_{\alpha\beta\sigma} - \bar{\Gamma}^\lambda_{\alpha\beta} \delta a_{\lambda\sigma} \, , \tag{5.62}$$

* For details on the functional space of the admissible test functions see end of Section 5.1.3.

which yields after some tedious, but straight forward manipulations

$$\delta S_{\alpha\beta\sigma} = \boldsymbol{a}_\sigma \cdot \delta \boldsymbol{r}_{|\alpha\beta} + \left[b_{\alpha\beta}\boldsymbol{n} + S^{\lambda}_{\alpha\beta}\boldsymbol{a}_\lambda\right] \cdot \delta \boldsymbol{a}_\sigma \ . \tag{5.63}$$

Note that we can rewrite the energy equivalently in terms of the contravariant form $S^{\sigma}_{\alpha\beta}$, which can be rewritten as

$$\delta S^{\sigma}_{\alpha\beta} = \boldsymbol{a}^\sigma \cdot \delta \boldsymbol{r}_{|\alpha\beta} + \left[b_{\alpha\beta}a^{\sigma\lambda}\boldsymbol{n} - S^{\lambda}_{\alpha\beta}\boldsymbol{a}^\sigma\right] \cdot \delta \boldsymbol{a}_\lambda \ , \tag{5.64}$$

see Steigmann [116] for details. The above redefinition of the variation of the different strain measures in terms of $\delta \boldsymbol{a}_\sigma = \delta \boldsymbol{r}_{,\sigma}$ and $\delta \boldsymbol{r}_{|\alpha\beta} = \delta \boldsymbol{r}_{,\alpha\beta} - \bar{\Gamma}^{\sigma}_{\alpha\beta}\delta \boldsymbol{r}_{,\sigma}$ allows us now to write the internal virtual work as

$$\delta \Psi = \boldsymbol{N}^\alpha \cdot \delta \boldsymbol{r}_{,\alpha} + \boldsymbol{M}^{\alpha\beta} \cdot \delta \boldsymbol{r}_{|\alpha\beta} \ . \tag{5.65}$$

Comparing (5.38) and (5.46) with (5.56), (5.61) and (5.63) yields

$$\begin{aligned} \boldsymbol{N}^\alpha = & \left[2\left(\frac{\partial \Psi}{\partial a_{\alpha\beta}}\right)^{\mathrm{sym}(\alpha\beta)} + \left(\frac{\partial \Psi}{\partial S_{\gamma\lambda\alpha}}\right)^{\mathrm{sym}(\gamma\lambda)} S^{\beta}_{\gamma\lambda}\right]\boldsymbol{a}_\beta + \\ & \left[\left(\frac{\partial \Psi}{\partial S_{\gamma\lambda\alpha}}\right)^{\mathrm{sym}(\gamma\lambda)} b_{\gamma\lambda} + \left(\frac{\partial \Psi}{\partial \kappa_{\gamma\lambda}}\right)^{\mathrm{sym}(\gamma\lambda)} S^{\alpha}_{\gamma\lambda}\right]\boldsymbol{n} \end{aligned} \tag{5.66}$$

and

$$\boldsymbol{M}^{\alpha\beta} = \left(\frac{\partial \Psi}{\partial S_{\alpha\beta\sigma}}\right)^{\mathrm{sym}(\alpha\beta)} \boldsymbol{a}_\sigma - \left(\frac{\partial \Psi}{\partial \kappa_{\alpha\beta}}\right)^{\mathrm{sym}(\alpha\beta)} \boldsymbol{n} \ . \tag{5.67}$$

Here, the symmetric part of the derivative $(\bullet)^{\mathrm{sym}_{\alpha\beta}}$ is introduced, where the symmetry condition is applied with respect to α and β, i.e.

$$\left(\frac{\partial \Psi}{\partial S_{\alpha\beta\sigma}}\right)^{\mathrm{sym}(\alpha\beta)} = \frac{1}{2}\left(\frac{\partial \Psi}{\partial S_{\alpha\beta\sigma}} + \frac{\partial \Psi}{\partial S_{\beta\alpha\sigma}}\right) \ . \tag{5.68}$$

Next, we resolve the second covariant derivative

$$\boldsymbol{M}^{\alpha\beta} \cdot \delta \boldsymbol{r}_{|\alpha\beta} = \boldsymbol{M}^{\alpha\beta} \cdot \delta \boldsymbol{r}_{,\alpha\beta} - \boldsymbol{M}^{\gamma\lambda}\bar{\Gamma}^{\alpha}_{\gamma\lambda} \cdot \delta \boldsymbol{r}_{,\alpha} \ , \tag{5.69}$$

such that we obtain for the virtual internal work

$$\delta \Psi = \tilde{\boldsymbol{N}}^\alpha \cdot \delta \boldsymbol{r}_{,\alpha} + \boldsymbol{M}^{\alpha\beta} \cdot \delta \boldsymbol{r}_{,\alpha\beta} \ , \tag{5.70}$$

where

$$\tilde{\boldsymbol{N}}^\alpha = \boldsymbol{N}^\alpha - \boldsymbol{M}^{\gamma\lambda}\bar{\Gamma}^{\alpha}_{\gamma\lambda} \ . \tag{5.71}$$

Insertion in (5.46) and subsequent integration by parts yields

$$\delta W^{\mathrm{int}} = \int_\Omega \left[(\tilde{\boldsymbol{N}}^\alpha - \boldsymbol{M}^{\alpha\beta}_{,\beta}) \cdot \delta \boldsymbol{r} + \boldsymbol{M}^{\alpha\beta} \cdot \delta \boldsymbol{r}_{,\beta}\right]_{,\alpha} - \left[\tilde{\boldsymbol{N}}^{\alpha}_{,\alpha} - \boldsymbol{M}^{\alpha\beta}_{,\alpha\beta}\right] \cdot \delta \boldsymbol{r}\, \mathrm{d}A \ . \tag{5.72}$$

The results of the divergence theorem applied on the first term can be decomposed in normal and tangential direction, i.e. normal and tangential to the edges of the shell

$$\int_{\partial\Omega} \Big[(\tilde{\boldsymbol{N}}^{\alpha} - \boldsymbol{M}^{\alpha\beta}_{,\beta}) \cdot \delta\boldsymbol{r} + \boldsymbol{M}^{\alpha\beta} \cdot \delta\boldsymbol{r}_{,\beta}\Big]\, \nu_{\alpha}\, \mathrm{d}S =$$
$$\int_{\partial\Omega} (\tilde{\boldsymbol{N}}^{\alpha} \cdot \delta\boldsymbol{r})\nu_{\alpha} - (\boldsymbol{M}^{\alpha\beta}_{,\beta} \cdot \delta\boldsymbol{r})\nu_{\alpha} + (\boldsymbol{M}^{\alpha\beta} \cdot \delta\boldsymbol{r}_{,\nu})\nu_{\alpha}\nu_{\beta} + (\boldsymbol{M}^{\alpha\beta} \cdot \delta\boldsymbol{r}_{,\tau})\nu_{\alpha}\tau_{\beta}\, \mathrm{d}S \tag{5.73}$$

where $\boldsymbol{\nu} = \nu_{\alpha}\boldsymbol{A}^{\alpha}$ is the rightward unit normal to $\partial\Omega$ and $\boldsymbol{\tau} = \tau_{\alpha}\boldsymbol{A}^{\alpha}$ the tangential unit vector. Moreover, $\delta\boldsymbol{r}_{,\nu} = \nu^{\alpha}\delta\boldsymbol{r}_{,\alpha}$ and $\delta\boldsymbol{r}_{,\tau} = \tau^{\alpha}\delta\boldsymbol{r}_{,\alpha}$ are the corresponding derivatives, decomposed in tangential and normal direction. A secondary integration by parts with subsequent application of the divergence theorem of the last term on the right hand side of (5.73) on the boundary $\partial\Omega$ of the shell yields

$$\int_{\partial\Omega} \boldsymbol{M}^{\alpha\beta}\nu_{\alpha}\tau_{\beta} \cdot \delta\boldsymbol{r}_{,\tau}\, \mathrm{d}S = \sum_{\partial^2\Omega} [\boldsymbol{M}^{\alpha\beta}\nu_{\alpha}\tau_{\beta}]_i \cdot \delta\boldsymbol{r}_i - \int_{\partial\Omega} (\boldsymbol{M}^{\alpha\beta}\nu_{\alpha}\tau_{\beta})' \cdot \delta\boldsymbol{r}\, \mathrm{d}S\,, \tag{5.74}$$

where $(\bullet)' = \mathrm{d}(\bullet)/\,\mathrm{d}S$ and the square brackets refer to the corner i at the boundary. Assuming that the principle of virtual work $0 = \delta W^{\text{int}} - \delta W^{\text{ext}}$ is valid with respect to the corresponding functional spaces of admissible solution and test functions defined at the end of this section, the external contribution can be formulated as

$$\delta W^{\text{ext}} = \int_{\Omega} \boldsymbol{g} \cdot \delta\boldsymbol{r}\, \mathrm{d}A + \int_{\Upsilon} \boldsymbol{t} \cdot \delta\boldsymbol{r} + \boldsymbol{\mu} \cdot \delta\boldsymbol{r}_{,\nu}\, \mathrm{d}S + \sum_i \boldsymbol{f}_i \cdot \delta\boldsymbol{r}_i\,. \tag{5.75}$$

where the edge Υ is assumed to be an open set on $\partial\Omega$ and the vertices $i \in \Xi = [1, .., n]$ are defined on $\partial^2\Omega$. Moreover, $\boldsymbol{g}$ denotes a distributed load, e.g. gravitational load. Eventually, we obtain the strong form of the second gradient problem as

$$(\tilde{\boldsymbol{N}}^{\alpha} - \boldsymbol{M}^{\alpha\beta}_{,\beta})_{,\alpha} + \boldsymbol{g} = \boldsymbol{0} \quad \text{on} \quad \Omega\,, \tag{5.76}$$

with boundary conditions at the edges

$$\begin{aligned} \boldsymbol{r} &= \tilde{\boldsymbol{r}} && \text{on} \quad \Upsilon_r\,, \\ \tilde{\boldsymbol{N}}^{\alpha}\nu_{\alpha} - \boldsymbol{M}^{\alpha\beta}_{,\beta}\nu_{\alpha} - (\boldsymbol{M}^{\alpha\beta}\nu_{\alpha}\tau_{\beta})' &= \boldsymbol{t} && \text{on} \quad \Upsilon_t\,, \\ \boldsymbol{r}_{,\nu} &= \tilde{\boldsymbol{k}} && \text{on} \quad \Upsilon_k\,, \\ \boldsymbol{M}^{\alpha\beta}\nu_{\alpha}\nu_{\beta} &= \boldsymbol{\mu} && \text{on} \quad \Upsilon_{\mu} \end{aligned} \tag{5.77}$$

and at the vertices

$$\begin{aligned} \boldsymbol{r}_i &= \tilde{\boldsymbol{r}}_i && \text{on} \quad \Xi_d\,, \\ \big[\boldsymbol{M}^{\alpha\beta}\nu_{\alpha}\tau_{\beta}\big]_i &= \boldsymbol{f}_i && \text{on} \quad \Xi_f \end{aligned} \tag{5.78}$$

of the shell. As usual for fourth-order boundary value problems, we decompose the whole boundary twice. First, $\Upsilon = \Upsilon_r \cup \Upsilon_t$, with respect to $\Upsilon_r \cap \Upsilon_t = \emptyset$, and second,

$\Upsilon = \Upsilon_k \cup \Upsilon_\mu$, with respect to $\Upsilon_k \cap \Upsilon_\mu = \emptyset$. Note that $\Upsilon_{r,t,k,\mu}$ are open sets on $\partial\Omega$, and we have additionally $\Xi = \Xi_f \cup \Xi_d$ with respect to $\Xi_f \cap \Xi_d = \emptyset$.

Eventually, we introduce the functional space of admissible solutions

$$\mathcal{S} = \{\boldsymbol{r} \in \mathcal{H}^2(\Omega) \,|\, \boldsymbol{r} = \tilde{\boldsymbol{r}} \text{ on } \Upsilon_r,\ \boldsymbol{r}_{,\nu} = \tilde{\boldsymbol{k}} \text{ on } \Upsilon_k,\ \boldsymbol{r}_i = \tilde{\boldsymbol{r}}_i \text{ on } \Xi_d\}\,, \tag{5.79}$$

and the space of admissible trial or test functions

$$\mathcal{V} = \{\delta\boldsymbol{r} \in \mathcal{H}^2(\Omega) \,|\, \delta\boldsymbol{r} = \boldsymbol{0} \text{ on } \Upsilon_r,\ \delta\boldsymbol{r}_{,\nu} = \boldsymbol{0} \text{ on } \Upsilon_k,\ \delta\boldsymbol{r}_i = \boldsymbol{0} \text{ on } \Xi_d\}\,, \tag{5.80}$$

required for the principle of virtual work. Note that the enforcement of the gradient terms in the space of admissible solutions is not trivial, see Schuß et al. [108] for details.

5.1.4 Spatial discretization

Due to the particular continuity requirements that follow from the incorporation of the curvature coefficients into the presented model, the isogeometric method is employed for the finite element analysis. Therefore, polynomial approximations of the deformed geometry in the shell mid-surface geometry $\boldsymbol{r}$ and its variation $\delta\boldsymbol{r}$ are defined as

$$\boldsymbol{r}^{\mathrm{h}} = \sum_{I\in\mathcal{I}} R^I \boldsymbol{q}_I \quad \text{and} \quad \delta\boldsymbol{r}^{\mathrm{h}} = \sum_{J\in\mathcal{I}} R^J \delta\boldsymbol{q}_J\,, \tag{5.81}$$

respectively, where $\boldsymbol{q}_I \in \mathbb{R}^3$ and $\delta\boldsymbol{q}_J \in \mathbb{R}^3$.

The approximations of the tangent vectors and their respective variations read

$$\boldsymbol{a}_\alpha^{\mathrm{h}} := \boldsymbol{r}_{,\alpha}^{\mathrm{h}} = \sum_{I\in\mathcal{I}} R_{,\alpha}^I \boldsymbol{q}_I \quad \text{and} \quad \delta\boldsymbol{a}_\alpha^{\mathrm{h}} := \delta\boldsymbol{r}_{,\alpha}^{\mathrm{h}} = \sum_{J\in\mathcal{J}} R_{,\alpha}^J \delta\boldsymbol{q}_J\,. \tag{5.82}$$

Thus, the approximations of the normal vector and its variation are given as

$$\boldsymbol{n}^{\mathrm{h}} := \frac{\boldsymbol{a}_1^{\mathrm{h}} \times \boldsymbol{a}_2^{\mathrm{h}}}{\|\boldsymbol{a}_1^{\mathrm{h}} \times \boldsymbol{a}_2^{\mathrm{h}}\|} \quad \text{and} \quad \delta\boldsymbol{n}^{\mathrm{h}} := \frac{(\boldsymbol{I} - \boldsymbol{n}^{\mathrm{h}} \otimes \boldsymbol{n}^{\mathrm{h}})(\boldsymbol{a}_1^{\mathrm{h}} \times \delta\boldsymbol{a}_2^{\mathrm{h}} + \delta\boldsymbol{a}_1^{\mathrm{h}} \times \boldsymbol{a}_2^{\mathrm{h}})}{\|\boldsymbol{a}_1^{\mathrm{h}} \times \boldsymbol{a}_2^{\mathrm{h}}\|}\,. \tag{5.83}$$

The approximations of the second covariant derivative and its variation read

$$\boldsymbol{a}_{\alpha,\beta}^{\mathrm{h}} := \boldsymbol{r}_{,\alpha\beta}^{\mathrm{h}} = \sum_{I\in\mathcal{I}} R_{,\alpha\beta}^I \boldsymbol{q}_I \quad \text{and} \quad \delta\boldsymbol{a}_{\alpha,\beta}^{\mathrm{h}} := \delta\boldsymbol{r}_{,\alpha\beta}^{\mathrm{h}} = \sum_{J\in\mathcal{J}} R_{,\alpha\beta}^J \delta\boldsymbol{q}_J\,. \tag{5.84}$$

Moreover, the membrane strains, the in- and out-of-plane curvature as well as their variations are discretized as

$$a_{\alpha\beta}^{\mathrm{h}} = \boldsymbol{a}_\alpha^{\mathrm{h}} \cdot \boldsymbol{a}_\beta^{\mathrm{h}} \quad \text{and} \quad \delta a_{\alpha\beta}^{\mathrm{h}} = \boldsymbol{a}_\alpha^{\mathrm{h}} \cdot \delta\boldsymbol{a}_\beta^{\mathrm{h}} + \boldsymbol{a}_\beta^{\mathrm{h}} \cdot \delta\boldsymbol{a}_\alpha^{\mathrm{h}}\,, \tag{5.85}$$

$$\kappa_{\alpha\beta}^{\mathrm{h}} = \boldsymbol{a}_{\alpha,\beta}^{\mathrm{h}} \cdot \boldsymbol{n}^{\mathrm{h}} \quad \text{and} \quad \delta\kappa_{\alpha\beta}^{\mathrm{h}} = \delta\boldsymbol{a}_{\alpha,\beta}^{\mathrm{h}} \cdot \boldsymbol{n}^{\mathrm{h}} + \boldsymbol{a}_{\alpha,\beta}^{\mathrm{h}} \cdot \delta\boldsymbol{n}^{\mathrm{h}}\,, \tag{5.86}$$

$$S^{\mathrm{h}}_{\alpha\beta\sigma} = \boldsymbol{a}^{\mathrm{h}}_{\alpha,\beta} \cdot \boldsymbol{a}^{\mathrm{h}}_{\sigma} - \bar{\Gamma}^{\mathrm{h},\lambda}_{\alpha\beta} a^{\mathrm{h}}_{\sigma\lambda} \quad \text{and} \quad \delta S^{\mathrm{h}}_{\alpha\beta\sigma} = \delta\boldsymbol{a}^{\mathrm{h}}_{\alpha,\beta} \cdot \boldsymbol{a}^{\mathrm{h}}_{\sigma} + \boldsymbol{a}^{\mathrm{h}}_{\alpha,\beta} \cdot \delta\boldsymbol{a}^{\mathrm{h}}_{\sigma} - \bar{\Gamma}^{\mathrm{h},\lambda}_{\alpha\beta} \delta a^{\mathrm{h}}_{\sigma\lambda}\,. \tag{5.87}$$

To complete the approximation of the strain measures, the discrete right Cauchy-Green tensor and its variation are given as follows

$$\hat{C}^{\mathrm{h}}_{\alpha\beta} = a^{\mathrm{h}}_{\alpha\beta} - 2\theta^3 \boldsymbol{a}^{\mathrm{h}}_{\alpha,\beta} \cdot \boldsymbol{n}^{\mathrm{h}} \quad \text{and} \quad \delta\hat{C}^{\mathrm{h}}_{\alpha\beta} = \delta a^{\mathrm{h}}_{\alpha\beta} + 2\theta^3 \delta\kappa^{\mathrm{h}}_{\alpha,\beta}\,. \tag{5.88}$$

Now, using above approximations the discrete version of the internal virtual work given in (5.46) reads

$$\delta W^{\mathrm{int,h}} = \int_{\Omega^{\mathrm{h}}} \Big[(n^{\alpha\beta}_{\mathrm{iso,h}} + n^{\alpha\beta}_{\mathrm{fib,h}}) \delta a^{\mathrm{h}}_{\alpha\beta} + (m^{\alpha\beta}_{\mathrm{iso,h}} + m^{\alpha\beta}_{\mathrm{fib,h}}) \delta\kappa^{\mathrm{h}}_{\alpha\beta} + m^{\alpha\beta\sigma}_{\mathrm{fib,h}} \delta S^{\mathrm{h}}_{\alpha\beta\sigma} \Big] \,\mathrm{d}A\,, \tag{5.89}$$

where discrete versions of the stress resultants for the matrix material are given as

$$n^{\alpha\beta}_{\mathrm{iso,h}} = \int_{\varpi_{\mathrm{iso}}} \frac{\partial \Psi^{\mathrm{iso,h}}(\hat{C}^{\mathrm{h}}_{ij})}{\partial \hat{C}^{\mathrm{h}}_{\alpha\beta}} \,\mathrm{d}\theta^3 \quad \text{and} \quad m^{\alpha\beta}_{\mathrm{iso,h}} = \int_{\varpi_{\mathrm{iso}}} 2\theta^3 \frac{\partial \Psi^{\mathrm{iso,h}}(\hat{C}^{\mathrm{h}}_{ij})}{\partial \hat{C}^{\mathrm{h}}_{\alpha\beta}} \,\mathrm{d}\theta^3 \tag{5.90}$$

and for the fiber material as

$$\begin{aligned} n^{\alpha\beta}_{\mathrm{fib,h}} &= \frac{\partial \Psi^{\mathrm{fib,h}}(a^{\mathrm{h}}_{\alpha\beta}, S^{\mathrm{h}}_{\alpha\beta\sigma}, \kappa^{\mathrm{h}}_{\alpha\beta})}{\partial a^{\mathrm{h}}_{\alpha\beta}}\,, \\ m^{\alpha\beta}_{\mathrm{fib,h}} &= \frac{\partial \Psi^{\mathrm{fib,h}}(a^{\mathrm{h}}_{\alpha\beta}, S^{\mathrm{h}}_{\alpha\beta\sigma}, \kappa^{\mathrm{h}}_{\alpha\beta})}{\partial \kappa^{\mathrm{h}}_{\alpha\beta}}\,, \\ m^{\alpha\beta\sigma}_{\mathrm{fib,h}} &= \frac{\partial \Psi^{\mathrm{fib,h}}(a^{\mathrm{h}}_{\alpha\beta}, S^{\mathrm{h}}_{\alpha\beta\sigma}, \kappa^{\mathrm{h}}_{\alpha\beta})}{\partial S^{\mathrm{h}}_{\alpha\beta\sigma}}\,. \end{aligned} \tag{5.91}$$

Moreover, the discrete version of the external virtual work given in (5.75) reads

$$\delta W^{\mathrm{ext,h}} = \int_{\Omega^{\mathrm{h}}} \boldsymbol{g} \cdot \delta\boldsymbol{r}^{\mathrm{h}} \,\mathrm{d}A + \int_{\Upsilon^{\mathrm{h}}} \boldsymbol{t} \cdot \delta\boldsymbol{r}^{\mathrm{h}} + \boldsymbol{\mu} \cdot (\nu^{\alpha,\mathrm{h}} \delta\boldsymbol{r}^{\mathrm{h}}_{,\alpha}) \,\mathrm{d}S + \sum_i \boldsymbol{f}_i \cdot \delta\boldsymbol{r}^{\mathrm{h}}_i\,, \tag{5.92}$$

where $\boldsymbol{\nu}^{\mathrm{h}} = \nu^{\alpha,\mathrm{h}} \boldsymbol{A}_{\alpha}$ is the discrete version of the rightward unit normal to $\partial\Omega^{\mathrm{h}}$.

5.1.5 Numerical examples

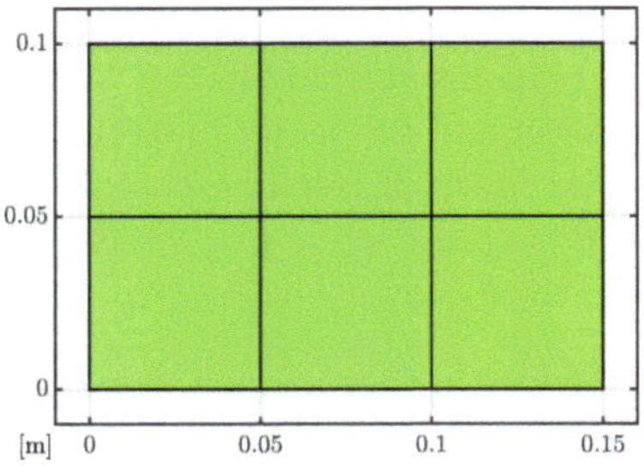

Figure 5.2: Computational mesh.

Incompressible matrix material	
Shear modulus μ	2 [MPa]
Fiber material	
Young's modulus E	6 [MPa]
Thickness h_{f}	0.001 [m]

Table 5.1: Material setting of the composite material.

The objective of this section is to present a series of numerical examples and to demonstrate the accuracy and performance of the proposed second gradient model. To begin, we examine a series of benchmark examples in terms of analytical solutions, verifying the comprehensive fiber material formulation.

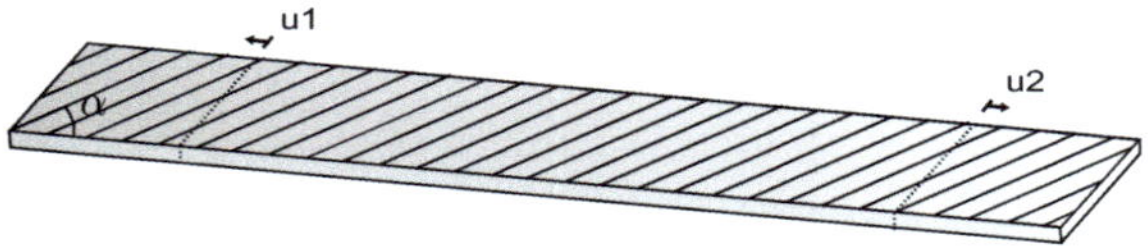

Figure 5.3: **Tensile test.** Problem setting. The lines illustrate the fiber structure.

Secondly, the combined matrix and fiber material is calibrated and validated with various experimental tests under different loading conditions and fiber orientations.

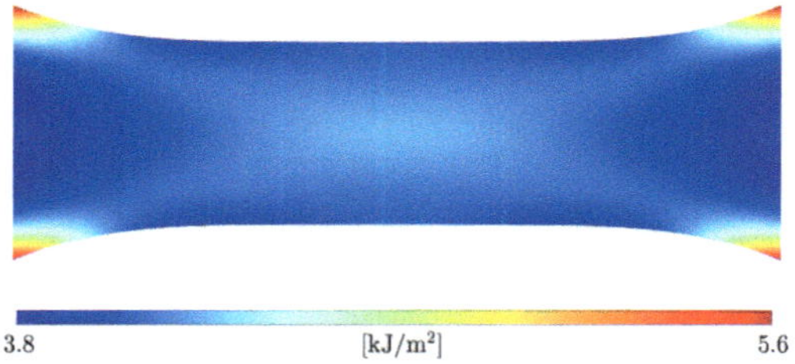

Figure 5.4: **Tensile test.** Strain energy density of the matrix material at displacement $u = 0.15\,[\mathrm{m}]$.

With the adjusted set of material parameters obtained in the previous examples, we examine a final example, demonstrating the application of the model to a more complex, non-planar geometry. In addition, the influence of the novel in-plane bending contributions of the fiber materials is particularly emphasized.

Verification

In a step by step verification of the numerical framework, we conduct several numerical tests which allow us to investigate each effect of the fiber material separately. Unless otherwise defined, a shell of size $l \times b = 0.15\,[\mathrm{m}] \times 0.1\,[\mathrm{m}]$ discretized by 3×2 cubic B-spline based elements is applied. The computational mesh and the material setting are given in Figure 5.2 and Table 5.1, respectively. Note that for each B-spline based mesh considered in this section, knot vectors of the structure $[\theta_1^\alpha = \ldots = \theta_{p_\alpha+1}^\alpha < \ldots < \theta_{\mathrm{n}_\alpha+1}^\alpha = \ldots = \theta_{\mathrm{n}_\alpha+p_\alpha+1}^\alpha]$ are applied.

Tensile test

We start our investigation with a tensile test to verify the behavior related to a stretching of the fibers, i.e. we set $a_1 = a_2 = h_\mathrm{f} E = 12\,[\mathrm{kN/m}]$ and neglect all other contributions

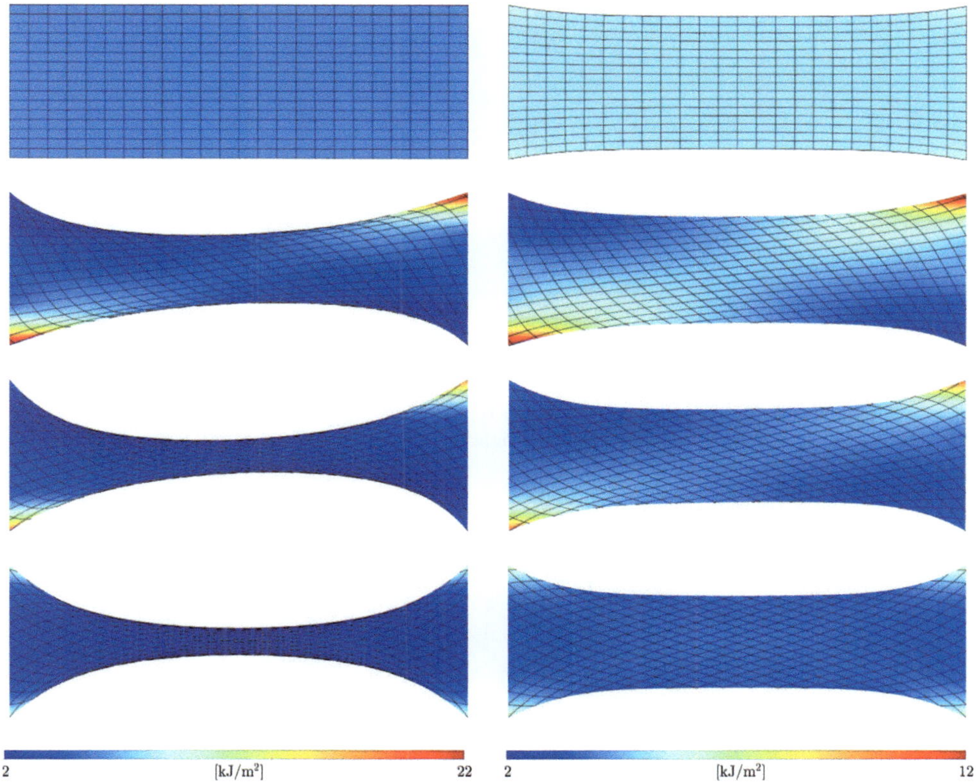

Figure 5.5: **Tensile test.** Strain energy density of the fiber material (left) and the composite material (right) at displacement $u = 0.15\,[\mathrm{m}]$ and fiber configurations of $\alpha = [0°, 15°, 30°, 45°]$ (from top to bottom. The lines illustrate the deformed fiber structure.

of the fiber material. The left edge of a shell is fixed and the right edge is moved in $\boldsymbol{e}_1$-direction by u, see Figure 5.3 for details.

In Figure 5.4, the strain energy per unit reference area is depicted for the pure matrix material using a shell thickness of $h = 0.002\,[\mathrm{m}]$. Moreover, strain energy densities for the pure fiber material with $h_\mathrm{f} = 0.002\,[\mathrm{m}]$ as well as for the composite material using the setting from Table 5.1 are shown in Figure 5.5. Therein fiber orientations of $\alpha = [0°, 15°, 30°, 45°]$ are applied. Note that the fiber material in the $0°$ configuration does not experience lateral strains under longitudinal displacement conditions at the right edge. Here, we obtain a constant energy density of $6\,[\mathrm{kJ/m^2}]$ which is in accordance with the analytical solution, i.e. $\sum_\alpha a_\alpha(\lambda_\alpha - 1)^2/2$ with $\lambda_1 = 2$ and $\lambda_2 = 1$.

Figure 5.6 shows the load deflection result, where results for the pure matrix material, the pure fiber material and the composite material are considered. Concerning the fiber

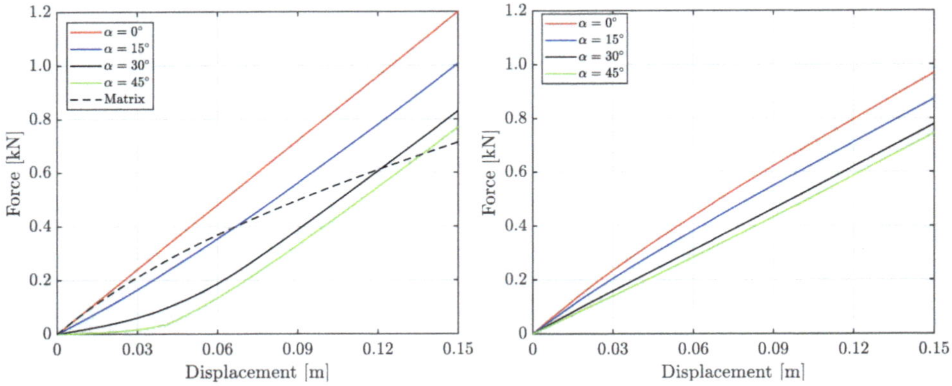

Figure 5.6: **Tensile test.** Load deflection result of the fiber and matrix material (left) and the composite material (right).

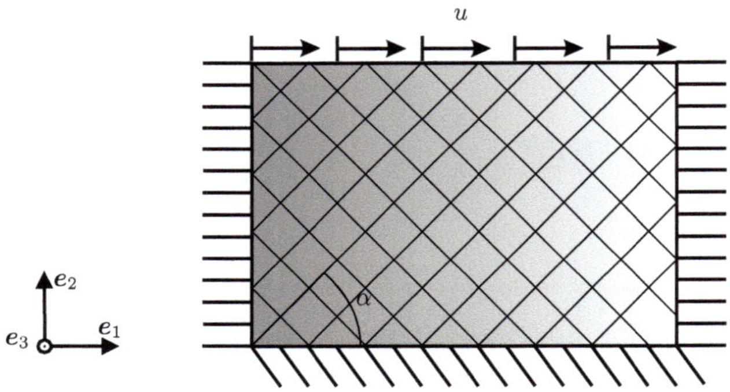

Figure 5.7: **Shear test.** Problem setting. The lines illustrate the fiber structure.

material with a 0° fiber orientation, we obtain a load of 1.2 [kN] which matches again with the analytical solution. This can be calculated in a straightforward manner using (5.66) and (5.77) by which we obtain a constant traction at the right boundary of $\boldsymbol{t} = \frac{a_1 u}{l} \boldsymbol{e}_1$.

Shear test

Next, we deal with a simple shear test. To be specific, the lower edge of a shell is fixed, the left and right edges are fixed in $\boldsymbol{e}_2$-direction and the upper edge is moved in $\boldsymbol{e}_1$-direction by u as illustrated in Figure 5.7. To verify the implementation, numerical results using pure fiber material ($h_\mathrm{f} = 0.002$ [m]) have to match analytical solutions. Therefore, we set $\alpha = 0°$, $a_1 = a_2 = 12$ [kN] and $a_3 = h_\mathrm{f}\mu = 4$ [kN/m]. All other parameters are set equal to zero.

Figure 5.8 and Figure 5.9 show the distribution of the strain energy density and the load deflection result, respectively. Therein, we obtain a homogeneous strain energy density of

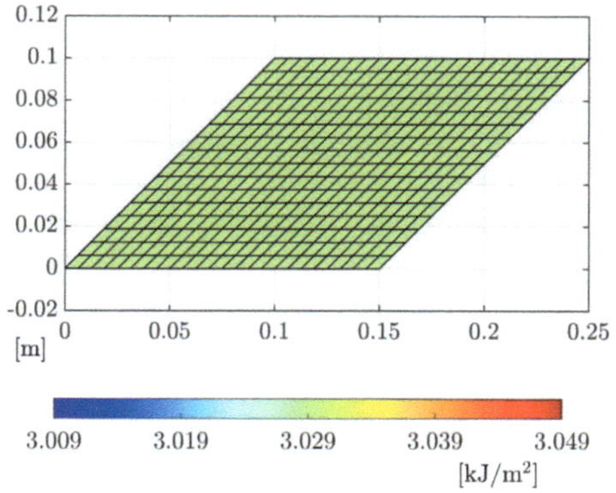

Figure 5.8: **Shear test.** Strain energy density of the fiber material at displacement $u = 0.1$ [m]. The lines illustrate the deformed fiber structure.

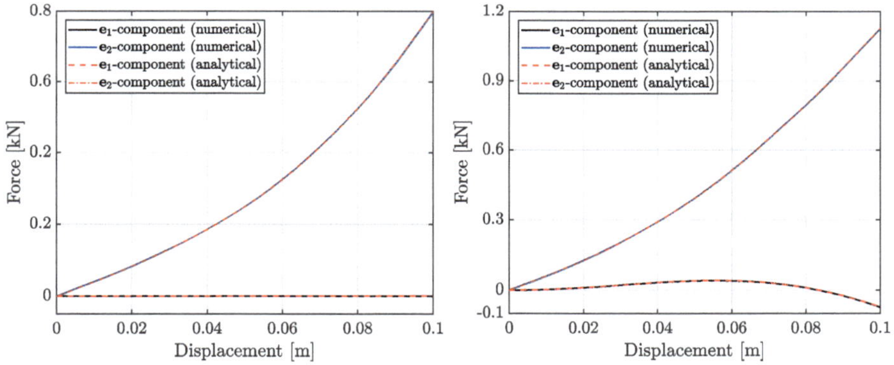

Figure 5.9: **Shear test.** Load deflection result. Numerical and analytical solutions of resultant forces at the right edge Υ_1 (left) and upper edge Υ_2 (right) are depicted.

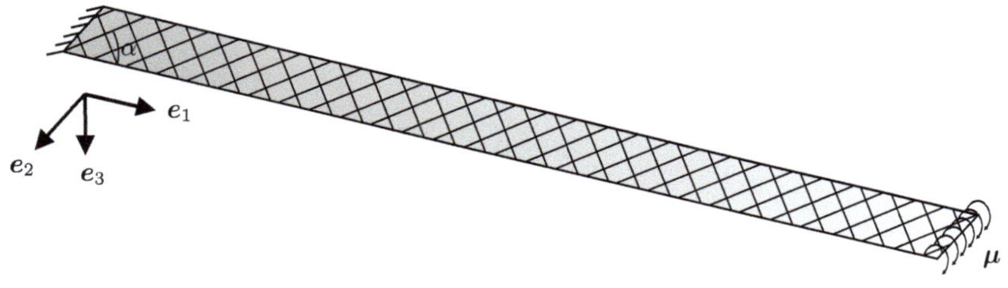

Figure 5.10: **Out-of-plane bending test.** Problem setting. The lines illustrate the fiber structure.

3029.3 [J/m^2]. The obtained results are in accordance to analytical solutions, which will be derived next.

Assuming a linear parametrization of the reference mid-surface

$$\bar{\boldsymbol{r}} = \theta^1 \boldsymbol{e}_1 + \theta^2 \boldsymbol{e}_2 \quad \text{with} \quad \theta^1 \in [0, l] \quad \text{and} \quad \theta^2 \in [0, b], \tag{5.93}$$

the analytical solution for pure fiber material with a fiber configuration of $\alpha = 0°$ can be derived as follows. Note that we obtain $\boldsymbol{A}_1 = \boldsymbol{e}_1$ and $\boldsymbol{A}_2 = \boldsymbol{e}_2$ due to (5.93). The isochoric deformed configuration of the mid-surface shown in Figure 5.8 is described by

$$\boldsymbol{r}(\theta^1, \theta^2) = \left(\theta^1 + \frac{\theta^2 u}{b}\right) \boldsymbol{e}_1 + \theta^2 \boldsymbol{e}_2, \tag{5.94}$$

where u is the predefined displacement of the upper boundary, see Figure 5.7. Now, applying (2.32) and (2.33) we obtain

$$\lambda_1 = 1, \quad \lambda_2 = \sqrt{\left(\frac{u}{b}\right)^2 + 1} \quad \text{and} \quad \varphi = \frac{\pi}{2} - \operatorname{acos}\left(\frac{\frac{u}{b}}{\sqrt{\left(\frac{u}{b}\right)^2 + 1}}\right) \tag{5.95}$$

as fiber stretches and the change of angle between the fibers, respectively. Insertion of the strain measures into (5.41) and (5.77) yields

$$\Psi^{\text{fib}} = \frac{a_2}{2}\left(\sqrt{\left(\frac{u}{b}\right)^2 + 1} - 1\right)^2 + \frac{a_3}{2} \tan\left(\frac{\pi}{2} - \operatorname{acos}\left(\frac{\frac{u}{b}}{\sqrt{\left(\frac{u}{b}\right)^2 + 1}}\right)\right)^2 \tag{5.96}$$

and after some straightforward calculations

$$\boldsymbol{N}^1 = a_3 \Pi \begin{bmatrix} 0 \\ 1 \\ 0 \end{bmatrix} \quad \text{and} \quad \boldsymbol{N}^2 = a_2 \left(1 - \frac{1}{\sqrt{\left(\frac{u}{b}\right)^2 + 1}}\right) \begin{bmatrix} \frac{u}{b} \\ 1 \\ 0 \end{bmatrix} + a_3 \frac{\Pi}{\left(\frac{u}{b}\right)^2 + 1} \begin{bmatrix} 1 \\ -\frac{u}{b} \\ 0 \end{bmatrix}, \tag{5.97}$$

where the abbreviation

$$\Pi = \frac{\tan\left(\frac{\pi}{2} - \operatorname{acos}\left(\frac{\frac{u}{b}}{\sqrt{\left(\frac{u}{b}\right)^2 + 1}}\right)\right)}{\cos\left(\frac{\pi}{2} - \operatorname{acos}\left(\frac{\frac{u}{b}}{\sqrt{\left(\frac{u}{b}\right)^2 + 1}}\right)\right)^2}. \tag{5.98}$$

is applied. Eventually, using (5.77) we can evaluate the resultant forces at the right boundary Υ_1 and upper boundary Υ_2 as

$$\boldsymbol{F}_{\Upsilon_1} = \int_{\Upsilon_1} \begin{bmatrix} \boldsymbol{N}^1 & \boldsymbol{N}^2 \end{bmatrix} \boldsymbol{\nu}_{\Upsilon_1} \, \mathrm{d}S = \boldsymbol{N}^1 b \quad \text{and} \quad \boldsymbol{F}_{\Upsilon_2} = \int_{\Upsilon_2} \begin{bmatrix} \boldsymbol{N}^1 & \boldsymbol{N}^2 \end{bmatrix} \boldsymbol{\nu}_{\Upsilon_2} \, \mathrm{d}S = \boldsymbol{N}^2 l \,, \tag{5.99}$$

respectively, where $\boldsymbol{\nu}_{\Upsilon_1} = [1, 0]^{\mathrm{T}}$ and $\boldsymbol{\nu}_{\Upsilon_2} = [0, 1]^{\mathrm{T}}$ are corresponding unit normal vectors. Thus, for the final deformation state, i.e. $u/b = 1$, we obtain

$$\Psi^{\mathrm{fib}}(u = 0.1\,[\mathrm{m}]) \approx 3029.3 \, \frac{\mathrm{J}}{\mathrm{m}^2} \tag{5.100}$$

as homogeneous strain energy distribution and the forces at the boundaries are

$$\boldsymbol{F}_{\Upsilon_1}(u = 0.1\,[\mathrm{m}]) = \begin{bmatrix} 0 \\ 800 \\ 0 \end{bmatrix} [\mathrm{N}] \quad \text{and} \quad \boldsymbol{F}_{\Upsilon_2}(u = 0.1\,[\mathrm{m}]) \approx \begin{bmatrix} -72.8 \\ 1127.2 \\ 0 \end{bmatrix} [\mathrm{N}] \,. \tag{5.101}$$

Out-of-plane bending test

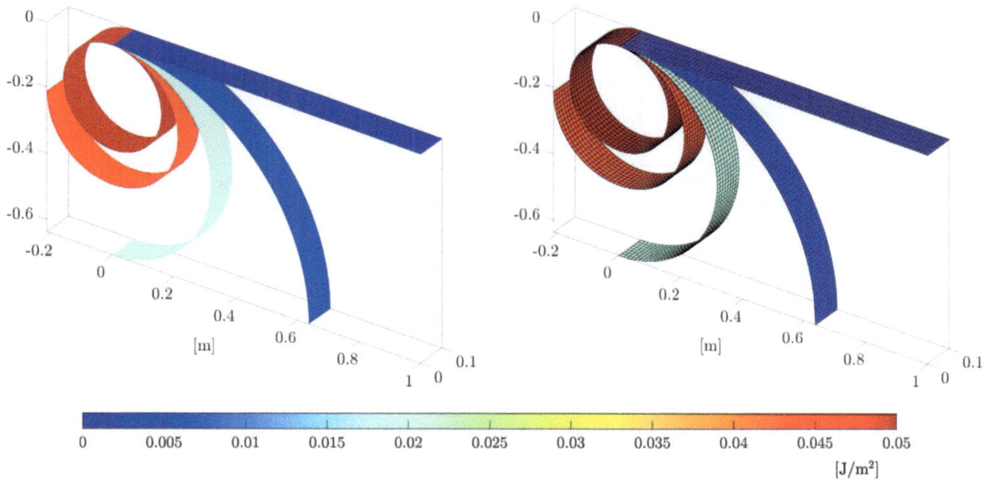

Figure 5.11: **Out-of-plane bending test.** Configuration of the shell and strain energy distribution at different load steps. Pure matrix material (left) and pure fiber material (right) are applied. The lines illustrate the deformed fiber structure.

For the out-of-plane bending test, the left edge of a shell is clamped and the right edge is subject to an external out-of-plane torque $\boldsymbol{\mu} = M\boldsymbol{e}_2$ as shown in Figure 5.10. In contrast to previous tests, the shell used for this test has a length of $l = 1\,[\mathrm{m}]$ and a

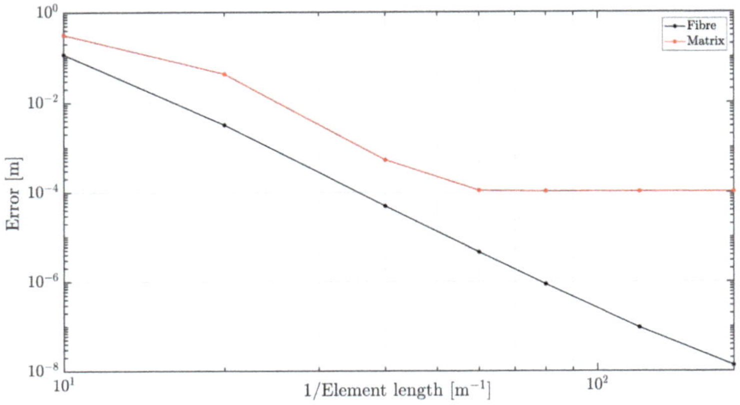

Figure 5.12: **Out-of-plane bending test.** Convergence results.

width of $b = 0.1$ [m] in the reference configuration, see Figure 5.11. Using a thickness of $h = 0.002$ [m] for pure matrix material and a thickness of $h_\mathrm{f} = 0.002$ [m] for pure fiber material, the area moment of inertia is obtained by $I = bh^3/12 = bh_\mathrm{f}^3/12 = 66.6667$ [mm^4]. The applied torque $M = 2\pi EI/l$ is chosen such that the shell describes a perfect circle in the asymptotic limit (see e.g. [69]) which holds for the fiber material by setting $\alpha = 0°$, $a_1 = h_\mathrm{f}E = 12$ [kN/m] and $k_1 = EI/b = 0.004$ [Nm]. Moreover, for the matrix material we apply the compressible model given in (5.40) with $\mu = E/2 = 3$ [MPa] and $\kappa = E/3 = 2$ [MPa] which corresponds to a Poisson ratio of $\nu = 0$.

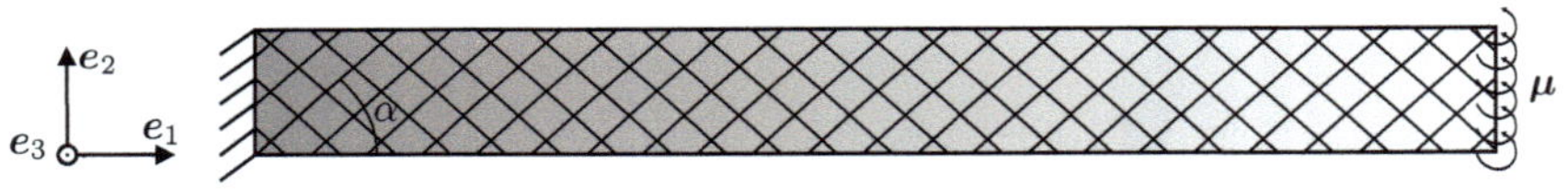

Figure 5.13: **In-plane bending test.** Problem setting. The lines illustrate the fiber structure.

In Figure 5.11, the configuration of the shell is plotted along with the strain energy distribution for pure matrix and pure fiber material at different load steps. The results shown therein are obtained by 80×8 cubic B-spline based elements. Moreover, Figure 5.12 shows the convergence properties for both materials, where the error is determined as gap between the ends of the shell. Here, we observe a limit of convergence at a value of approximately 10^{-4} [m] for pure matrix material, whereas for pure fiber material an error of less than $2 \cdot 10^{-8}$ [m] is obtained. Note that the limit of convergence using pure matrix material can be justified by known locking effects of Kirchhoff-Love shell elements as it may occur even for high polynomial approximations, see e.g. [66, 12, 54, 42, 14]. In addition, the usage of an internal Newton-Raphson iteration within the implementation of a compressible neo-Hookean model for the matrix material also limits the convergence.

In-plane bending test

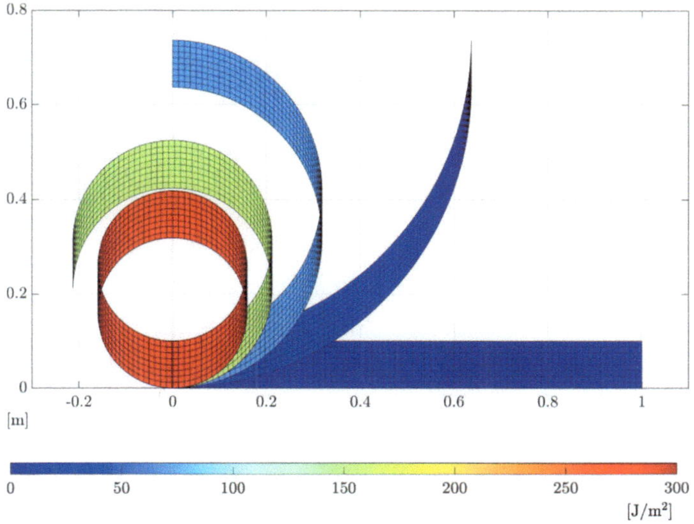

Figure 5.14: **In-plane bending test.** Configuration of the shell and strain energy distribution at different load steps using pure fiber material. The lines illustrate the deformed fiber structure.

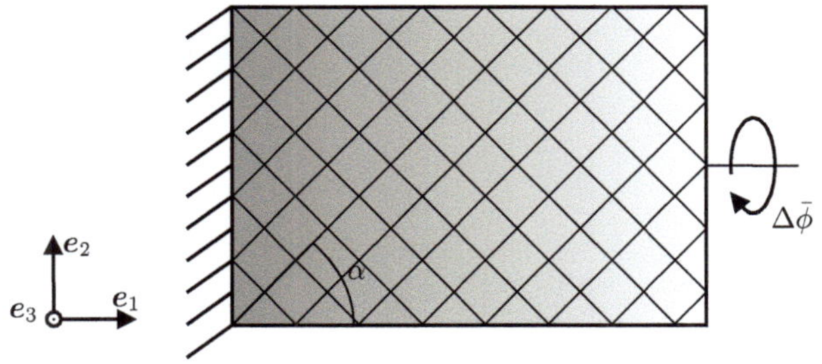

Figure 5.15: **Torsion test.** Problem setting. The lines illustrate the fiber structure.

Next, a verification of the in-plane bending behavior of the fiber material is demonstrated. Therefore, we use the shell mesh defined for the out-of-plane bending test and set $\alpha = 0°$, $a_1 = h_\mathrm{f}E = 12\,[\mathrm{kN/m}]$ with $h_\mathrm{f} = 0.002\,[\mathrm{m}]$ and $g_1 = g_2 = E\bar{I}/b = 10\,[\mathrm{Nm}]$ with $\bar{I} = h_\mathrm{f}b^3/12 = 166667\,[\mathrm{mm}^4]$. All other parameters of the fiber material are set equal to zero. Moreover, we apply $\boldsymbol{\mu} = M\boldsymbol{e}_3$ as an external in-plane bending torque, see Figure 5.13 for details. Again, the applied torque $M = 2\pi E\bar{I}/l$ is chosen such that the shell describes a perfect circle in the asymptotic limit, which is demonstrated in Figure 5.14. Even if this result is not physically reasonable for a real composite due to the material penetration, it verifies the implementation related to in-plane bending of pure fiber material.

Torsion test

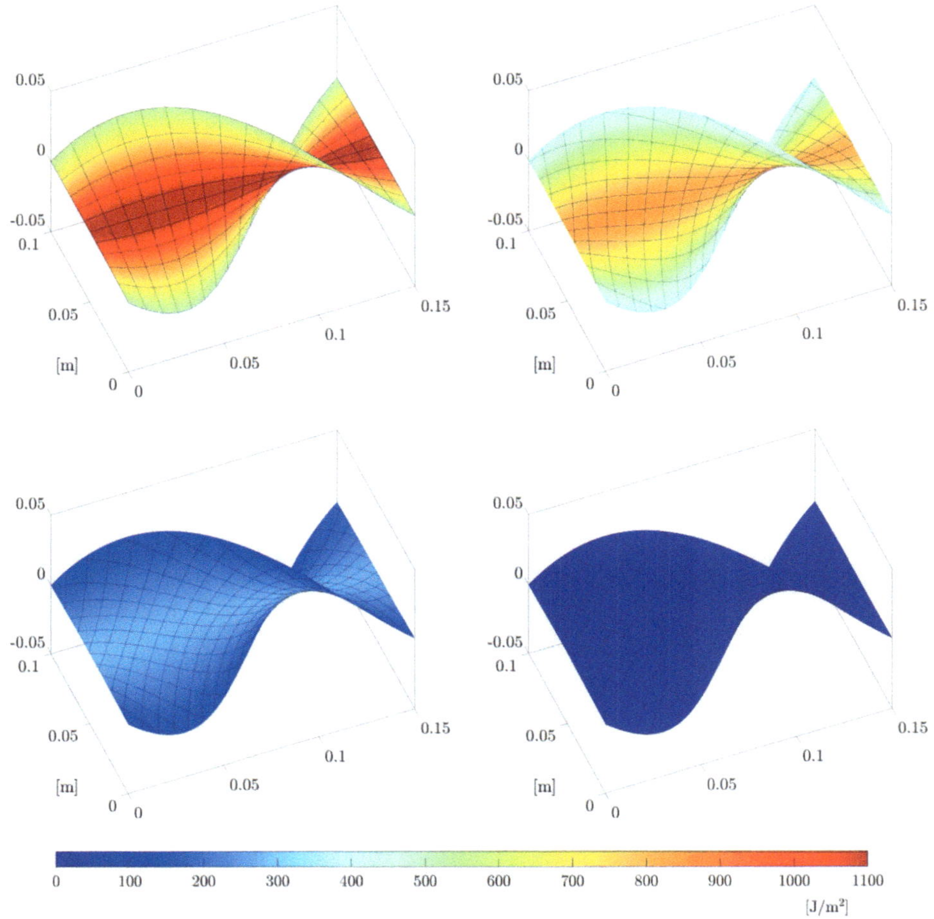

Figure 5.16: **Torsion test.** Strain energy distribution of the fiber material at $\Delta\bar{\phi} = 180°$ using different fiber configurations $\alpha = [0°, 15°, 30°, 45°]$ (from left to right and top to bottom). The lines illustrate the deformed fiber structure.

Eventually, we verify the numerical framework related to a torsional deformation of the fiber material, see Figure 5.15. Therefore, the deformation is predetermined such that the shell undergoes a constant twist of $\partial\bar{\phi}/\partial X_1 = 1200\,[°/\mathrm{m}]$ with $X_1 = \boldsymbol{X} \cdot \boldsymbol{e}_1$. Moreover, the parameters of the fiber material are defined by $h_\mathrm{f} = 0.002\,[\mathrm{m}]$ and $k_3 = \mu I_\mathrm{p}/b = 5.002\,[\mathrm{Nm}]$ with a polar moment of inertia $I_\mathrm{p} = b h_\mathrm{f}(b^2 + h_\mathrm{f}^2)/12 = 166733\,[\mathrm{mm}^4]$. Again, all other parameters are set equal to zero.

Strain energy densities obtained at $\Delta\bar{\phi} = 180°$ are depicted in Figure 5.16 for different fiber configurations. As expected, we obtain a highly stressed region along the central axis of the shell especially in the $0°$ fiber configuration, whereas the fiber material does not

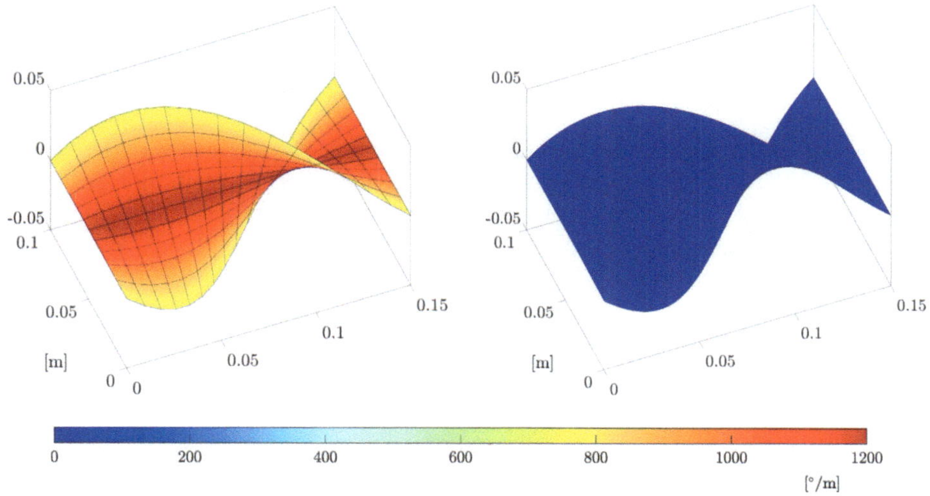

Figure 5.17: **Torsion test.** Twist of fiber at $\Delta\bar{\phi} = 180°$ using a fiber configuration of $\alpha = 0°$ (left) and $45°$ (right). The lines illustrate the deformed fiber structure.

contribute to the strain energy in the 45° configuration since the fibers are not twisted. This is also evident in Figure 5.17, where the twist of fiber is plotted for fiber configurations of $\alpha = 0°$ and $\alpha = 45°$. Concerning the 0° configuration, fibers in $\boldsymbol{e}_1$-direction undergo a twist of $\partial\bar{\phi}/\partial X_1 = 1200\,[°/\mathrm{m}]$ at the central axis of the shell. Moreover, analytical details related to the torsional test will be detailed next, verifying the numerical results shown in Figure 5.16.

Assuming a linear parametrization of the reference mid-surface, cf. (5.93), with $\theta^1 \in [0, l]$ and $\theta^2 \in [-b/2, b/2]$ such that $\boldsymbol{A}_1 = \boldsymbol{e}_1$ and $\boldsymbol{A}_2 = \boldsymbol{e}_2$, strain energy distributions for pure fiber material with different fiber configurations can be evaluated as follows. The predefined deformation of the mid-surface shown in Figure 5.16 and 5.17, respectively, is described by

$$\boldsymbol{r}(\theta^1, \theta^2) = \theta^1 \boldsymbol{e}_1 + \theta^2 \cos\left(\frac{\pi\theta^1}{l}\right) \boldsymbol{e}_2 + \theta^2 \sin\left(\frac{\pi\theta^1}{l}\right) \boldsymbol{e}_3 \,. \tag{5.102}$$

Applying (2.32), $(2.35)_2$ and $(2.51)_2$ we obtain

$$\kappa_{11} = \kappa_{22} = 0 \quad \text{and} \quad \kappa_{12} = \kappa_{21} = -\frac{\pi}{\sqrt{(\theta^2\pi)^2 + l^2}} \,. \tag{5.103}$$

Concerning different fiber configurations defined by α, the vector fields $\boldsymbol{L}$ and $\boldsymbol{M}$ in the reference configuration are given by

$$L^1 = \cos(\alpha), \quad L^2 = \sin(\alpha), \quad M^1 = -\sin(\alpha), \quad M^2 = \cos(\alpha) \,. \tag{5.104}$$

Eventually, insertion of these derivations into (5.41) yields

$$\Psi^{\text{fib}} = \frac{k_3}{2}(K_3)^2 = \frac{k_3}{2}\frac{\pi^2}{(\theta^2\pi)^2 + l^2}(\cos(\alpha)^2 - \sin(\alpha)^2)^2 \,. \tag{5.105}$$

Note that for the evaluation all material parameters are set equal to zero excepted k_3. Obviously, for the $\alpha = 45°$ fiber configuration, the strain energy distribution related to a torsional deformation of the shell is identical to zero, whereas for each other configuration, the highest value is obtained at the central axis of the shell, i.e. at $\theta^2 = 0$. We obtain for the $\alpha = 0°$ fiber configuration

$$\Psi^{\text{fib}}(\theta^2 = 0) \approx 1097.06\,\frac{\text{J}}{\text{m}^2} \quad \text{and} \quad \Psi^{\text{fib}}(\theta^2 = b/2) = \Psi^{\text{fib}}(\theta^2 = -b/2) \approx 523.25\,\frac{\text{J}}{\text{m}^2}\,, \tag{5.106}$$

for the $\alpha = 15°$ fiber configuration

$$\Psi^{\text{fib}}(\theta^2 = 0) \approx 822.8\,\frac{\text{J}}{\text{m}^2} \quad \text{and} \quad \Psi^{\text{fib}}(\theta^2 = b/2) = \Psi^{\text{fib}}(\theta^2 = -b/2) \approx 392.44\,\frac{\text{J}}{\text{m}^2} \tag{5.107}$$

and for the $\alpha = 30°$ fiber configuration

$$\Psi^{\text{fib}}(\theta^2 = 0) \approx 274.27\,\frac{\text{J}}{\text{m}^2} \quad \text{and} \quad \Psi^{\text{fib}}(\theta^2 = b/2) = \Psi^{\text{fib}}(\theta^2 = -b/2) \approx 130.81\,\frac{\text{J}}{\text{m}^2}\,. \tag{5.108}$$

Calibration and validation

We investigate Tepex®dynalite 102-RG600(1)/47 from LANXESS as prototypical composite material, consisting of 47% vol. of a woven fabric (roving glass) and polycaprolactam (PA 6) as matrix material. The matrix material is hydrophilic with a maximum absorption of moisture between 2.6% and 3.4%. For the investigations presented here, a single layer material with layer thickness of 0.5 [mm] has been specifically manufactured for this test, since multi-layered material with different fiber angles intermix various physical effects within the measurements.

Figure 5.18: **Tension test.** Clamping devise with 36 mm specimen, undeformed configuration.

The surrounding matrix material stiffens the woven fabric and connects the junctions of the fibers. The fibers within the composite material have significant bending stiffness due to the matrix material surrounding every single string within the fibers. Additional stiffening mechanisms are present depending on the actual weave of the woven fabric, since the fibers are subject to frictional behavior at the junctions. These effects arise only for the combined composite material and thus, cannot be separated within the experimental investigations, i.e. we always obtain a mean value of the material characteristics.

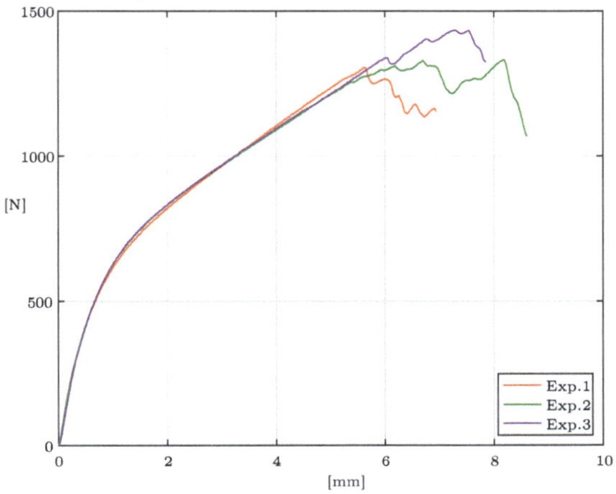

Figure 5.19: **Tension test.** Experimental force-displacement curves for 25 [mm] specimen with 30° fiber orientation.

The matrix material is in general governed by a far more complex constitutive behavior in the inelastic regime. Moreover, fibers and matrix might separate in the presence of large deformations, i.e. both materials can be characterized by separately assigned deformation fields, coupled via corresponding constitutive laws. For now, we restrict ourself to the hyperelastic regime without fiber separation.

Tension test

We first elaborated tension tests with 0°, 15°, 30° and 45° fiber orientation. Here, experimental data of 25 [mm] specimen are compared with numerical results using 34×10 cubic elements. Shear and bulk modulus used within the compressible model (5.40) for the PA 6 matrix material are assumed to be $\mu = 384.62$ [N/mm^2] and $\kappa = 833.33$ [N/mm^2], respectively. This setting corresponds to Young's modulus of $E_{\text{iso}} = 1000$ [N/mm^2] and Poisson ratio of $\nu = 0.3$. Moreover, for the glass fibers we assume $E_{\text{fib}} = 73000$ [N/mm^2]. The thickness of the material is $h = 0.5$ [mm], with 47% vol. of woven fabric, from which 50% are aligned in L and M direction, respectively. Thus, we obtain the material constants for the fiber stretch contributions via $0.5hE_{\text{fib}} \times 0.47$ as $a_{1/2} = 8577.5$ [N/mm], which corresponds to a combined Young's modulus of 17685 [N/mm^2]. As can be seen in Figure 5.20 for the 0° fiber orientation, the simulation fits perfectly the experiments.

The value for $a_3 = 250$ [N/mm] is related to the interaction between the fibers and can not be derived from other constants. Thus, we have fitted the value to the experimental

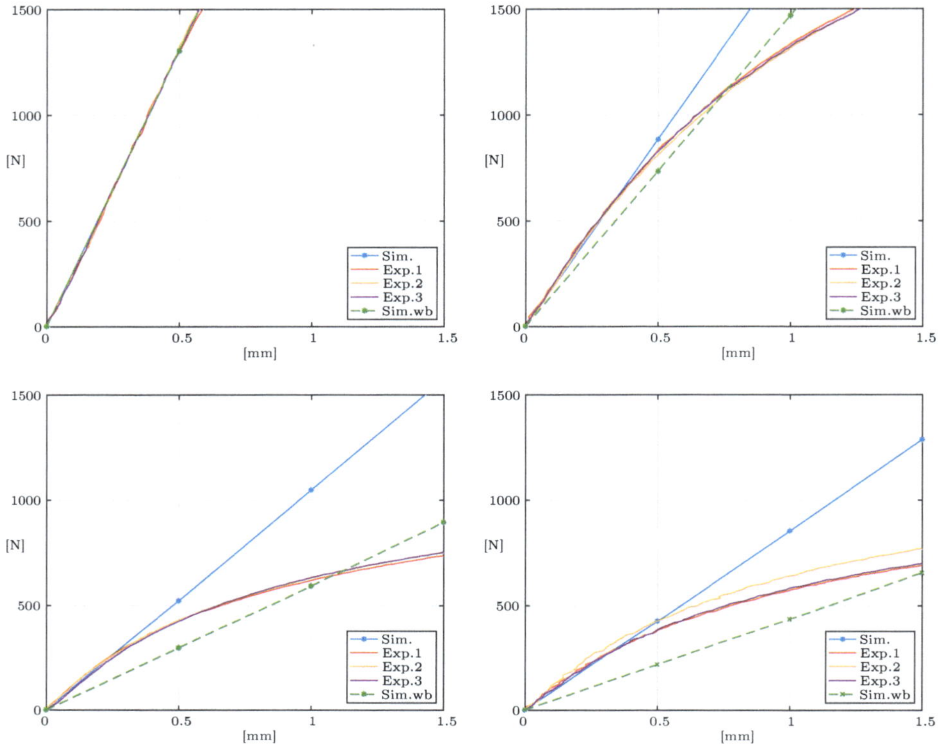

Figure 5.20: **Tension test.** Force-displacement curves of the tension test for 25 [mm] specimen with 0°, 15°, 30° and 45° fiber orientation (left to right, top to down). Simulation results are compared with experimental investigations. Additionally, simulation results without in-plane bending stiffness (Sim. wb) are shown to highlight the global effect of the non-classical in-plane curvature term.

observation obtained by a fiber orientation of 45°, see Figure 5.20 (right, bottom). The bending stiffness $g_1 = g_2$ on the other hand is related to Young's modulus via $g_{1/2} = E_{\text{fib}}\bar{I}/b = 893500$ [Nmm], where the second moment of inertia is $\bar{I} = b^3 h_{\text{f}}/12$ with $h_{\text{f}} = 0.47h$. Note that this approach is only a motivation for choosing the material constants g_1 and g_2, since this allows us a comparison with analytical results as shown in Section 5.1.5. The values $k_{1/2/3}$ do not contribute in these tension tests and g_3 was set to zero. Additional simulation results presented in Figure 5.20 demonstrate the effects of the in-plane bending term, i.e. we set $g_1 = g_2 = 0$, to compare the results with an anisotropic Kirchhoff-Love shell formulation neglecting the higher-gradient terms.

This example shows that the bending stiffness has considerable effects on the global stiffness and is not negligible. The constitutive parameters chosen with respect to the different matrix and fiber materials and the respective geometry allow for a nearly perfect

Figure 5.21: **Bending test.** Deformed configuration.

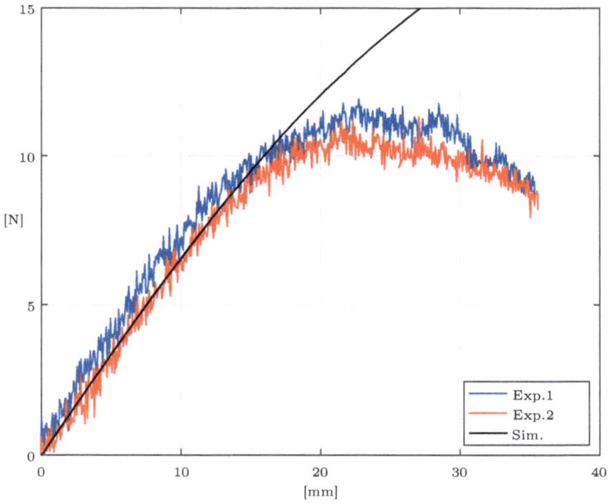

Figure 5.22: **Bending test.** Force-displacement curves of the out-of-plane bending test for a 25 [mm] specimen with 0° fiber orientation. Simulation results are compared with experimental investigations.

prediction of the macroscopic behavior of this material with microstructure. Only the parameter a_3 is fitted, since the shear stiffness of the fibers depends on the weave of the fabric and the production process in terms of applied pressure, heat and cooling rate. This result illustrates the necessity to include higher-gradient terms in the model for

Figure 5.23: **Bending test.** Permanent deformation of the sample after a full loading cycle.

fiber-reinforced polymers.

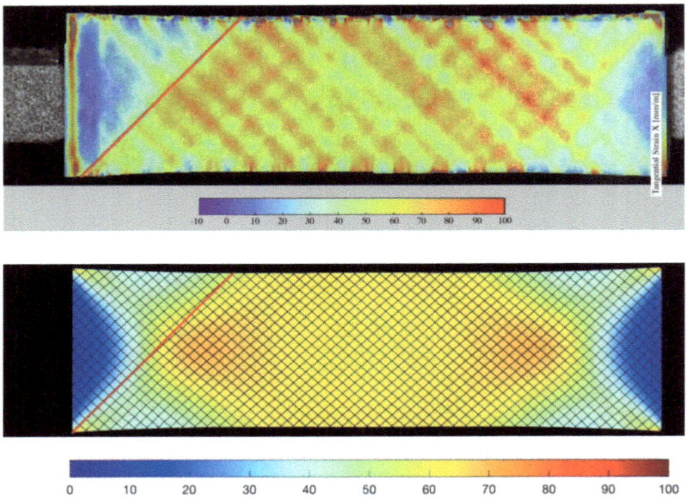

Figure 5.24: **Tension test.** DIC measurement of a 25 [mm] specimen in 45° configuration (top) and simulation results (bottom) at 4.6 [mm] total displacement, colors indicate local stretches in horizontal direction in [mm/m]. Black lines indicate the fiber direction in the current configuration.

Bending test

Next, we investigate the out-of-plane bending stiffness, hence we validate the values for k_1 and k_2. Therefore, a 4-point bending test as shown in Figure 5.21 is conducted. For the experiments, a stack of 5 specimen with 125 [mm] $\times$ 25 [mm] in 0° fiber configuration were used, such that we expect a stiffness of 5 times $k_{1/2} = E_{\text{fib}} I/b = 78.949$ [Nmm], where $I = bh_{\text{f}}^3/12$ and $h_{\text{f}} = 0.47h$. The support structure has a width of 40 [mm] between the inner brackets, and 30 [mm] between the outer and the inner brackets, such that the whole device has a width of 100 [mm].

For the numerical validation, we discretize the shell with 50 $\times$ 10 cubic elements with prescribed displacements in upward direction at the contact points of the support structure, such that the shell can slide over the bracket points and the system is subject to pure bending. The results in Figure 5.22 demonstrate the consistency of the formulation with the experimental observations until a displacement of about 15 [mm]. Afterwards, inelastic effects can be observed, resulting in a permanent deformation of the sample as can be observed in Figure 5.23. After a displacement of 25 [mm], the sample starts to slip over the brackets.

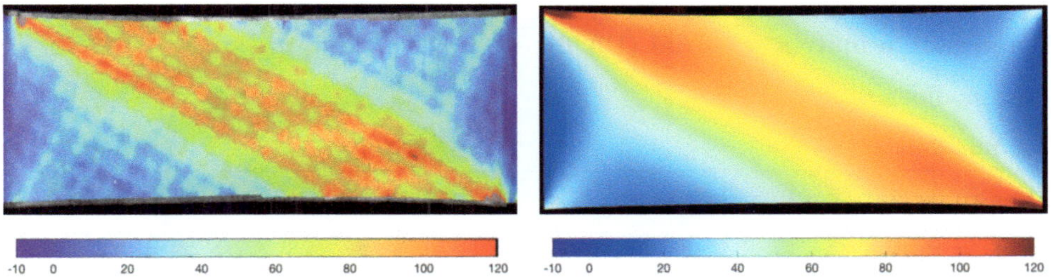

Figure 5.25: **Tension test.** DIC measurement of a 36 [mm] specimen in 60° configuration (top) and simulation results (bottom) at 4.6 [mm] total displacement, colors indicate local stretches in horizontal direction in [mm/m].

Digital Image correlation

Finally, we show in Figure 5.24 (top) the results of the Digital image correlation for the 25 [mm] specimen with 45° fiber orientation and a total displacement on the left boundary of 4.6 [mm]. Note that both sides are clamped in such a way that the displacement is prescribed and, additionally, the higher-order boundary conditions are also set, preventing in-plane twist and torsion of the fibers. A direct comparison with simulation results is not feasible, since the current model does not incorporate elastic softening or inelastic effects of the polymeric matrix material. However, we want to highlight the strain-gradient effects in terms of the resulting deformation patterns. Therefore, the stiffness of the matrix material E_{iso} and the bending stiffness g_1 and g_2 is reduced by a factor of 500 to fit the globally reduced stiffness of the inelastic deformed composite. The bending stiffness of the pure fiber is nearly zero, but contribute considerably to the global stiffness if the fibers are embedded within the matrix material. If the matrix is subject to inelastic effects, the bending stiffness is effected as well. Note that we have observed pronounced inelastic deformations by evaluating hysteresis curves. A detailed investigation of inelastic effects including plastification of the matrix material will follow in the next chapter. In both plots in Figure 5.24, the strain-gradient effects result in the stiffness of the fibers against in-plane bending, which can be observed at the crosspoint of the fibers as highlighted by the red line.

Using the same material setting, we compare next results for a 36 [mm] specimen in 30/60° configuration, see Figure 5.25. As can be seen, the deformation patterns fit in general, although the fibers at the vertices of the sample already start to fracture.

As final result of this section, we summarize the determined and validated material setting of our prototypical composite material Tepex®dynalite 102-RG600(1)/47 in Table 5.2.

Warp and weft direction are given in 0° and 90° direction, respectively. Both exhibits a slightly different material behavior and are not perfectly aligned due to the production process. This can be calibrated using different values for $a_{1/2}$ and $g_{1/2}$, which we omit here since the general behavior of this second-gradient material is the key issue in this paper.

Compressible matrix material (PA 6)	
Shear modulus μ	384.62 [N/mm^2]
Bulk modulus κ	833.33 [N/mm^2]
Fiber material (roving glass)	
Tensile stiffness a_1, a_2	8577.5 [N/mm]
Shear stiffness a_3	250 [N/mm]
In-plane bending stiffness g_1, g_2	893500 [Nmm]
Out-of-plane bending stiffness k_1, k_2	78.949 [Nmm]

Table 5.2: Material setting of Tepex®dynalite 102-RG600(1)/47.

Shaft example

This last example demonstrates the applicability of the proposed model to complex geometries with non-planar reference configurations. Therefore we consider a shaft of length 50 [mm] and radius $50/\pi$ [mm] as shown in Figure 5.26, using the validated material setting given in Table 5.2. Additionally, we set $g_3 = 0$ and $k_3 = 0$. The discretization of the shaft consists of 200 cubic B-spline based elements. To obtain a closed cylinder, the extended Mortar method [108] is applied between the opposing edges in peripheral direction of the shell, maintaining G^1-continuity, see also Dittmann et al. [40] for more details. The displacement on the left in $\boldsymbol{e}_2$-direction of the cylinder is fixed in space, whereas the displacement on the right is fixed on the $(\boldsymbol{e}_1, \boldsymbol{e}_3)$-plane and rotated as illustrated in Figure

5.26. Note that no higher-order boundary conditions according to $(5.77)_3$ and (5.78) are applied.

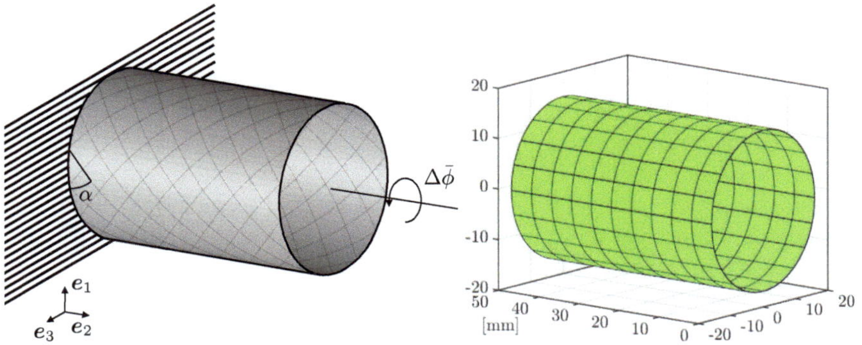

Figure 5.26: **Shaft example.** Problem setting and computational mesh.

First, we investigate a pure matrix material. After a prescribed angle of $\Delta\bar{\phi} = 3.3°$, the shell starts to buckle. The last state of equilibrium is obtained at an angle of $\Delta\bar{\phi} = 12.6°$. Note that e.g. an arc length method can be used to calculate further states in the geometrically unstable regime. In Figure 5.28, the applied torque necessary to obtain the prescribed twist at the right hand side of the shell versus the angle of twist $\Delta\bar{\phi}$ is plotted for the pure matrix as well as for the composite material with different fiber orientations with and without the in-plane bending terms.

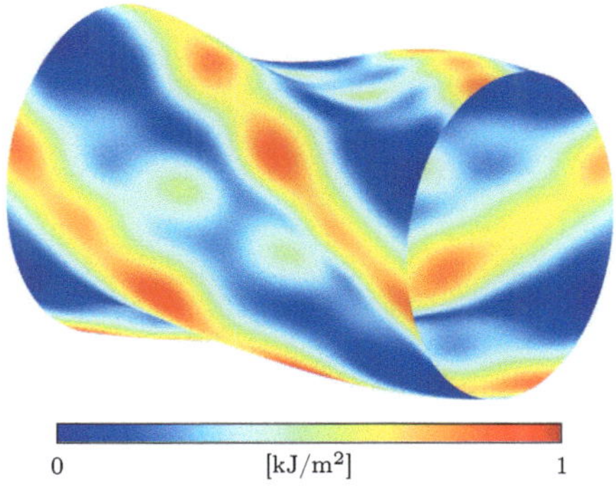

Figure 5.27: **Shaft example.** Strain energy density of the matrix material.

Next, the composite material is investigated. In Figure 5.30, top left, the strain energy for a 0° fiber configuration after a prescribed twist of $\Delta\bar{\phi} = 90°$ is displayed. Note that the fiber orientation and not the computational mesh is shown. The corresponding torque is given again in Figure 5.28. Several issues can be highlighted here. First, the higher-order boundary conditions as presented in $(5.77)_3$ are not set and thus, the fibers can twist in-plane at the boundary. This can be observed very well in Figure 5.30, top left. If bending would be prohibited at the boundary, the fibers have to remain orthogonal with regard to the boundary. Since these additional boundary conditions are unknown to a first gradient theory or even to the classical Kirchhoff-Love shell theory which only considers the out- and not the in-plane contributions, the proposed formulation is the only way to enforce this consistently within the mechanical formulation. Second, the in-plane terms stabilize the shell against buckling. Figure 5.31, top left, shows the results if the additional in-plane terms are not present, i.e. the material constants g_1 and g_2 are set to zero. Thus, we obtain the results for an anisotropic Kirchhoff-Love formulation, usual fitted via a first-gradient homogenization process, showing large buckling modes. To illustrate this further, the strain energy of the in-plane contributions for the deformed configuration (cf. Figure 5.31, top left) is plotted in Figure 5.29. As can be observed, a dramatic increase of the in-plane strain energy occurs for this heavily deformed configuration, hence, in the mechanical equilibrium this buckling mode is not possible.

Additionally, results for 15°, 30° and 45° fiber orientation are plotted (top right, lower left and lower right picture in Figure 5.30 and 5.31). The corresponding torque is plotted again in Figure 5.28. Notably, a nearly constant strain energy distribution can be observed in Figure 5.30, lower right and top left.

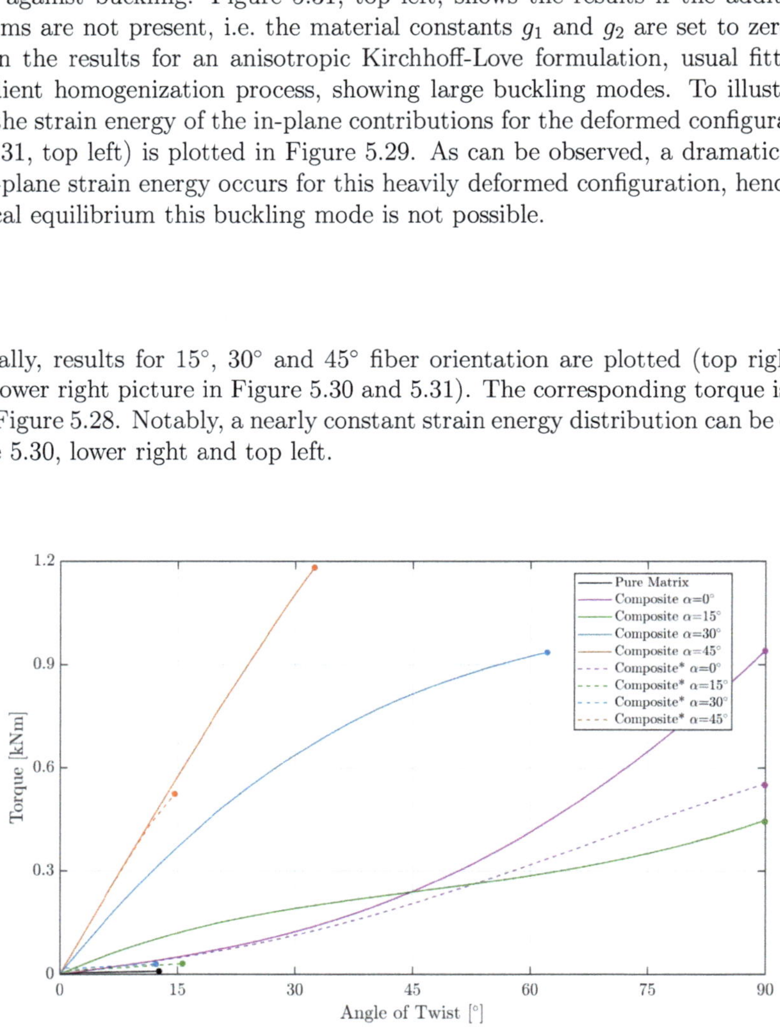

Figure 5.28: **Shaft example.** Torque versus angle of twist. Displayed are the results for the pure matrix material, the composite and the composite without the additional in-plane contributions (composite*).

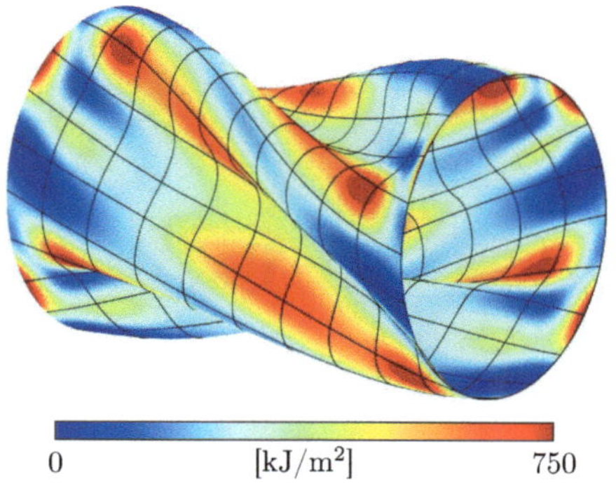

Figure 5.29: **Shaft example.** Estimated strain energy density of the in-plane contribution using the deformed configuration in 5.31, top left.

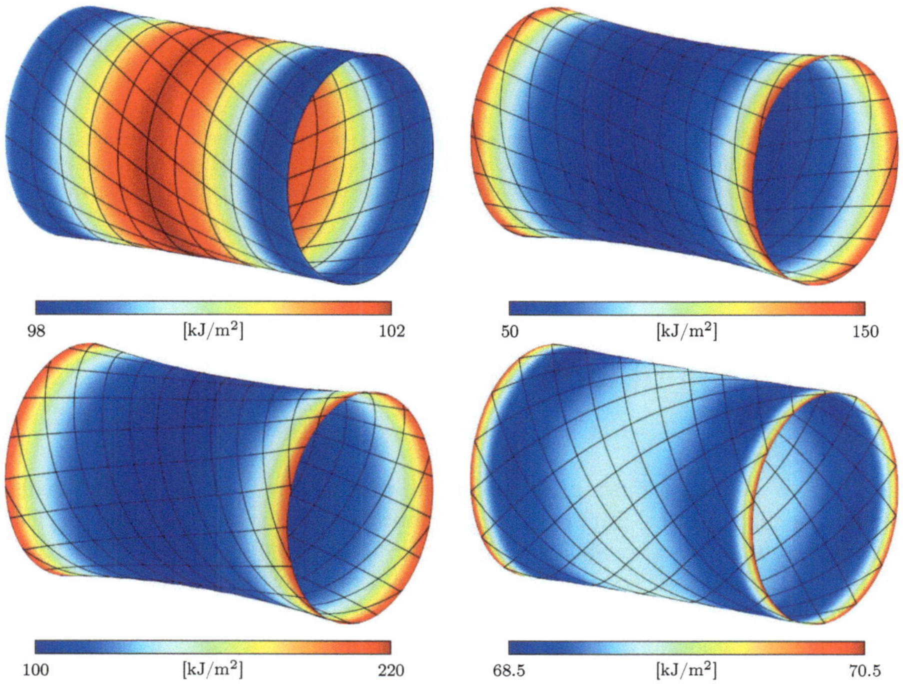

Figure 5.30: **Shaft example.** Strain energy density of the composite material with different fiber orientations $\alpha = [0°, 15°, 30°, 45°]$ (from left to right and top to bottom) plotted at the final deformation state marked in Figure 5.28.

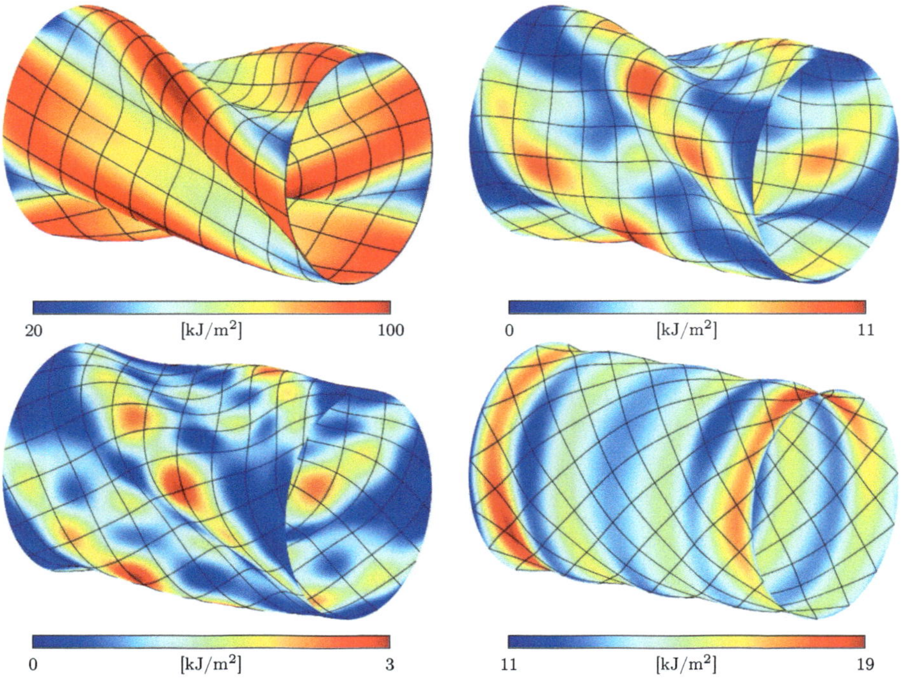

Figure 5.31: **Shaft example.** Strain energy density of the composite material with different fiber orientations $\alpha = [0°, 15°, 30°, 45°]$ (from left to right and top to bottom) without in-plane contributions plotted at the final deformation state marked in Figure 5.28.

5.2 Solid element formulation

In the previous section, we used experimental results to demonstrate that our shell formulation with higher-order in-plane bending contributions, in contrast to the classical Cauchy continuum, allows for independent modeling of the composite without the need to recalibrate the material setting for specific fiber orientation. In this section, this generalized continuum formulation for long fiber materials is adapted to a more general three-dimensional solid element, in a more straightforward manner. As in the previous section, we focus on modeling of the mechanical behavior of the fibers, and use a standard elastic matrix material. To avoid being too repetitive, we will skip over the spatial discretization, which will be discussed in depth in the following chapter. The implementation is then verified by numerical examples, focusing on the validating of the bending contributions.

5.2.1 Configuration and kinematics

As introduced in Chapter 2 in the context of classical continuum mechanics, we consider a body of interest $\mathcal{B}_0 \in \mathbb{R}^3$ in its reference configuration along with the deformation mapping $\varphi(\boldsymbol{X}, t)$. The deformation gradient $\boldsymbol{F}(\boldsymbol{X}, t)$ with $J := \det[\boldsymbol{F}] > 0$ can also be defined in terms of the principal stretches λ_a with $a = \{1, 2, 3\}$ and the principal directions of the left and right stretch tensors $\boldsymbol{n}_a$ and $\boldsymbol{N}_a$ as

$$\boldsymbol{F} = \sum_a \lambda_a \boldsymbol{n}_a \otimes \boldsymbol{N}_a \quad \text{and} \quad J = \prod_a \lambda_a. \tag{5.109}$$

Since we intend to extend the matrix material to elastoplasticity in the next chapter, which relies on different mechanisms for the deviatoric and volumetric contributions, we introduce the isochoric components of the principal strains

$$\bar{\lambda}_a = (J)^{-1/3} \lambda_a = \prod_b (\lambda_b)^{-1/3} \lambda_a, \tag{5.110}$$

and define the isochoric deformation gradient as

$$\bar{\boldsymbol{F}} = \sum_a \bar{\lambda}_a \boldsymbol{n}_a \otimes \boldsymbol{N}_a. \tag{5.111}$$

Given that these deformation measurements adequately characterize a typical elastic matrix material, we now focus on the fiber material.

Discussions on higher-order contributions for the constitutive modeling of composites with flexural resistance have previously been addressed, i.e. in Asmanoglo and Menzel [10, 11], using the framework as provided in the preliminary work of Spencer and Soldatos [114] and Soldatos [113]. Since the formulation is in terms of the reference configuration, the non-orthogonality of the fibers in the actual configuration has no effect as long as the reference configuration is orthogonal.

A completely different but very straightforward and physically illustrative approach was presented by dell'Isola [31]. Therein a detailed investigation of a Piola homogenization of a discrete system of extensional and rotational 2D springs is investigated. We will derive the exact same deformation measures within a strain-gradient continuum framework and as a general extension to 3D, and show that it can be recast into the formulation of Asmanoglo and Menzel [10, 11].

As with the shells, we introduce two fiber threads oriented in the $\boldsymbol{L}$ and $\boldsymbol{M}$ directions, respectively, where $\boldsymbol{L}$ and $\boldsymbol{M}$ are constant orthogonal unit vector fields within the body in the reference configuration. We then derive the stretch of the respective fiber

$$\lambda_{\mathrm{L}} = \|\boldsymbol{l}\| = \|\boldsymbol{F}\boldsymbol{L}\| \quad \text{and} \quad \lambda_{\mathrm{M}} = \|\boldsymbol{m}\| = \|\boldsymbol{F}\boldsymbol{M}\|. \tag{5.112}$$

With the normalized fiber directions

$$\tilde{\boldsymbol{l}} = \frac{\boldsymbol{l}}{\lambda_{\mathrm{L}}} = \frac{\boldsymbol{F}\boldsymbol{L}}{\|\boldsymbol{F}\boldsymbol{L}\|} \quad \text{and} \quad \tilde{\boldsymbol{m}} = \frac{\boldsymbol{m}}{\lambda_{\mathrm{M}}} = \frac{\boldsymbol{F}\boldsymbol{M}}{\|\boldsymbol{F}\boldsymbol{M}\|} \tag{5.113}$$

we then define the change of the angle between both fiber threads as

$$\begin{aligned}\varphi &= \operatorname{acos}(\tilde{\boldsymbol{l}} \cdot \tilde{\boldsymbol{m}}) - \frac{\pi}{2} \\ &= \operatorname{acos}\left(\frac{(\boldsymbol{FL}) \cdot (\boldsymbol{FM})}{\|\boldsymbol{FL}\| \|\boldsymbol{FM}\|}\right) - \frac{\pi}{2}.\end{aligned} \tag{5.114}$$

Eventually the corresponding measure for the shear deformation of the fiber material yields

$$\phi = \tan(\varphi). \tag{5.115}$$

Within the context of generalized continua, specifically the strain-gradient theory, the fiber bending is described taking into account the gradients of the deformed fiber vectors, i.e. $\nabla \boldsymbol{l} = \nabla \boldsymbol{FL}$ and $\nabla \boldsymbol{m} = \nabla \boldsymbol{FM}$. In particular, we consider

$$\nabla \boldsymbol{l} \boldsymbol{L} = \lambda_{\mathrm{L}} \nabla \tilde{\boldsymbol{l}} \boldsymbol{L} + (\nabla \lambda_{\mathrm{L}} \cdot \boldsymbol{L}) \tilde{\boldsymbol{l}} \quad \text{and} \quad \nabla \boldsymbol{m} \boldsymbol{M} = \lambda_{\mathrm{M}} \nabla \tilde{\boldsymbol{m}} \boldsymbol{M} + (\nabla \lambda_{\mathrm{M}} \cdot \boldsymbol{M}) \tilde{\boldsymbol{m}} \tag{5.116}$$

which are projections of the fiber configuration gradients onto the initial fiber direction, cf. Asmanoglo and Menzel [10]. These expressions include terms related to stretch gradients of the fibers as well as fiber curvatures. We introduce the curvature measure for the fiber initially aligned in $\boldsymbol{L}$-direction as

$$\begin{aligned}\boldsymbol{\kappa}_{\mathrm{L}} &= \frac{1}{\lambda_{\mathrm{L}}} \nabla \tilde{\boldsymbol{l}} \boldsymbol{L} \\ &= \frac{1}{\lambda_{\mathrm{L}}^2} (\nabla \boldsymbol{l} - \tilde{\boldsymbol{l}} \otimes \nabla \lambda_{\mathrm{L}}) \boldsymbol{L} \\ &= \frac{1}{\|\boldsymbol{FL}\|^2} \left(\nabla \boldsymbol{FL} - \frac{\boldsymbol{FL}}{\|\boldsymbol{FL}\|} \otimes \left(\frac{\boldsymbol{FL}}{\|\boldsymbol{FL}\|} \otimes \boldsymbol{L} \right) : \nabla \boldsymbol{F} \right) \boldsymbol{L}\end{aligned} \tag{5.117}$$

and for the fiber initially aligned in $\boldsymbol{M}$-direction as

$$\begin{aligned}\boldsymbol{\kappa}_{\mathrm{M}} &= \frac{1}{\lambda_{\mathrm{M}}} \nabla \tilde{\boldsymbol{m}} \boldsymbol{M} \\ &= \frac{1}{\lambda_{\mathrm{M}}^2} (\nabla \boldsymbol{m} - \tilde{\boldsymbol{m}} \otimes \nabla \lambda_{\mathrm{M}}) \boldsymbol{M} \\ &= \frac{1}{\|\boldsymbol{FM}\|^2} \left(\nabla \boldsymbol{FM} - \frac{\boldsymbol{FM}}{\|\boldsymbol{FM}\|} \otimes \left(\frac{\boldsymbol{FM}}{\|\boldsymbol{FM}\|} \otimes \boldsymbol{M} \right) : \nabla \boldsymbol{F} \right) \boldsymbol{M}.\end{aligned} \tag{5.118}$$

5.2.2 Variational formulation

Energetic response

The stored elastic energy density of the composite material is defined by the functional

$$\Psi = \zeta \Psi_{\mathrm{mat}}(\bar{\boldsymbol{F}}, J) + \frac{1-\zeta}{2} \Psi_{\mathrm{fib}}(\lambda_{\mathrm{L}}, \lambda_{\mathrm{M}}, \phi, \boldsymbol{\kappa}_{\mathrm{L}}, \boldsymbol{\kappa}_{\mathrm{M}}) \tag{5.119}$$

where $\zeta \in [0,1]$ is the volume fraction of the matrix material. As a representative matrix material, we use the modified volumetric-isochoric decoupled Ogden material model

$$\begin{aligned}\Psi_{\text{mat}} &= \Psi_{\text{mat}}^{\text{iso}}\left(\bar{\boldsymbol{F}}(\lambda_1,\lambda_2,\lambda_3)\right) &+ \;\; &\Psi_{\text{mat}}^{\text{vol}}\left(J(\lambda_1,\lambda_2,\lambda_3)\right) \\ &= \sum_a\sum_b \frac{\mu_b}{\alpha_b}\left((\bar{\lambda}_a)^{\alpha_b}-1\right) &+ \;\; &\frac{\kappa}{\beta^2}\left(\beta\ln(J)+(J)^{-\beta}-1\right).\end{aligned} \tag{5.120}$$

The parameters μ_b and α_b with $b=\{1,\ldots,N\}$ are related to the shear modulus and the parameters κ and β are related to the bulk modulus. Assuming that the fiber portion in both directions is identical, the corresponding contribution of the fiber material is defined by

$$\begin{aligned}\Psi_{\text{fib}} &= \frac{1}{2}a\left((\lambda_{\text{L}}-1)^2+(\lambda_{\text{M}}-1)^2\right)+b\,\phi^2 \\ &+\frac{1}{2}\left(\boldsymbol{\kappa}_{\text{L}}\cdot\boldsymbol{c}\,\boldsymbol{\kappa}_{\text{L}}+\boldsymbol{\kappa}_{\text{M}}\cdot\boldsymbol{c}\,\boldsymbol{\kappa}_{\text{M}}\right)\end{aligned} \tag{5.121}$$

where a and b are stiffness parameters related to stretch and shear of the fiber material. Moreover, the stiffness tensor related to fiber curvature is given as

$$\boldsymbol{c} = c_{\#}(\boldsymbol{l}\otimes\boldsymbol{l}+\boldsymbol{m}\otimes\boldsymbol{m})+c_{\perp}\boldsymbol{n}\otimes\boldsymbol{n} \quad \text{with} \quad \boldsymbol{n}=\boldsymbol{l}\times\boldsymbol{m}. \tag{5.122}$$

Previous works such as Asmanoglo and Menzel [10], used a scalar bending parameter, resulting in an isodirectional bending stiffness and thus limited modeling capabilities. In the proposed tensorial argument $\boldsymbol{c}$, the different stiffness parameters $c_{\#}$ and $c_{\perp}$ can be interpreted as the in-plane and out-of-plane bending stiffness, respectively. These can in turn account for the geometric dependence, namely the respective plane moment of inertia of the fiber bundle, as detailled in Section 5.1.5. Note that we assume the same cross section for both fiber threads.

Regarding the partial derivatives therein, we introduce first relations related to the Kirchhoff stress $\boldsymbol{\tau}=\boldsymbol{\tau}_{\text{mat}}+\boldsymbol{\tau}_{\text{fib}}$ as

$$\begin{aligned}\boldsymbol{\tau}_{\text{mat}} &= \boldsymbol{\tau}_{\text{mat}}^{\text{dev}}+\boldsymbol{\tau}_{\text{mat}}^{\text{vol}} \\ &= \sum_a\left(\tau_{\text{mat},a}^{\text{dev}}+\tau_{\text{mat},a}^{\text{vol}}\right)\boldsymbol{n}_a\otimes\boldsymbol{n}_a \\ &= \zeta\sum_a\lambda_a\left(\frac{\partial\Psi_{\text{mat}}^{\text{iso}}}{\partial\lambda_a}+\frac{\partial\Psi_{\text{mat}}^{\text{vol}}}{\partial\lambda_a}\right)\boldsymbol{n}_a\otimes\boldsymbol{n}_a\end{aligned} \tag{5.123}$$

and

$$\boldsymbol{\tau}_{\text{fib}} = \frac{1-\zeta}{2}\left(\frac{\partial\Psi_{\text{fib}}}{\partial\lambda_{\text{L}}}\frac{\partial\lambda_{\text{L}}}{\partial\boldsymbol{F}}+\frac{\partial\Psi_{\text{fib}}}{\partial\lambda_{\text{M}}}\frac{\partial\lambda_{\text{M}}}{\partial\boldsymbol{F}}+\frac{\partial\Psi_{\text{fib}}}{\partial\phi}\frac{\partial\phi}{\partial\boldsymbol{F}}+\frac{\partial\Psi_{\text{fib}}}{\partial\boldsymbol{\kappa}_{\text{L}}}\frac{\partial\boldsymbol{\kappa}_{\text{L}}}{\partial\boldsymbol{F}}+\frac{\partial\Psi_{\text{fib}}}{\partial\boldsymbol{\kappa}_{\text{M}}}\frac{\partial\boldsymbol{\kappa}_{\text{M}}}{\partial\boldsymbol{F}}\right)\boldsymbol{F}^{\text{T}}, \tag{5.124}$$

the higher-order stress of the fiber materials as

$$\mathfrak{P}_{\text{fib}} = \frac{1-\zeta}{2}\left(\frac{\partial\Psi_{\text{fib}}}{\partial\boldsymbol{\kappa}_{\text{L}}}\frac{\partial\boldsymbol{\kappa}_{\text{L}}}{\partial\nabla\boldsymbol{F}}+\frac{\partial\Psi_{\text{fib}}}{\partial\boldsymbol{\kappa}_{\text{M}}}\frac{\partial\boldsymbol{\kappa}_{\text{M}}}{\partial\nabla\boldsymbol{F}}\right). \tag{5.125}$$

Strong Form

To identify and collect the internal contributions to the boundaries of the mechanical field, i.e. the resulting bending moments and normal stress contributions, we derive the internal virtual work as

$$\delta W^{\text{int}} = \int_{\mathcal{B}_0} \delta\Psi_{\text{mat}} + \delta\Psi_{\text{fib}} \, \mathrm{d}V = \int_{\mathcal{B}_0} \boldsymbol{\tau} : \nabla_x \delta\boldsymbol{\varphi} + \mathfrak{P} \,\vdots\, \nabla\nabla\delta\boldsymbol{\varphi} \, \mathrm{d}V. \tag{5.126}$$

The usage of the transformation

$$\boldsymbol{\tau} : \nabla_x \delta\boldsymbol{\varphi} = \boldsymbol{\tau}\boldsymbol{F}^{-\mathrm{T}} : \nabla\delta\boldsymbol{\varphi} \tag{5.127}$$

along with integration by parts yields

$$\delta W^{\text{int}} = \int_{\mathcal{B}_0} -\nabla\cdot(\boldsymbol{\tau}\boldsymbol{F}^{-\mathrm{T}})\cdot\delta\boldsymbol{\varphi} + \nabla\cdot(\delta\boldsymbol{\varphi}\cdot\boldsymbol{\tau}\boldsymbol{F}^{-\mathrm{T}}) - \nabla\cdot\mathfrak{P} : \nabla\delta\varphi + \nabla\cdot(\nabla\delta\boldsymbol{\varphi} : \mathfrak{P}) \, \mathrm{d}V. \tag{5.128}$$

A second integration by parts related to the third term yields

$$\delta W^{\text{int}} = \int_{\mathcal{B}_0} \nabla\cdot(\nabla\cdot\mathfrak{P} - \boldsymbol{\tau}\boldsymbol{F}^{-\mathrm{T}})\cdot\delta\boldsymbol{\varphi} + \nabla\cdot(\delta\boldsymbol{\varphi}\cdot(\boldsymbol{\tau}\boldsymbol{F}^{-\mathrm{T}} - \nabla\cdot\mathfrak{P})) + \nabla\cdot(\nabla\delta\boldsymbol{\varphi} : \mathfrak{P}) \, \mathrm{d}V. \tag{5.129}$$

In a last step, we apply the divergence theorem for the second and third term such that we obtain

$$\delta W^{\text{int}} = \int_{\mathcal{B}_0} \nabla\cdot(\nabla\cdot\mathfrak{P} - \boldsymbol{\tau}\boldsymbol{F}^{-\mathrm{T}})\cdot\delta\boldsymbol{\varphi} \, \mathrm{d}V + \int_{\partial\mathcal{B}_0} ((\boldsymbol{\tau}\boldsymbol{F}^{-\mathrm{T}} - \nabla\cdot\mathfrak{P})\boldsymbol{N})\cdot\delta\boldsymbol{\varphi} + \mathfrak{P}\boldsymbol{N} : \nabla\delta\boldsymbol{\varphi} \, \mathrm{d}A. \tag{5.130}$$

Note that also contributions in tangential direction at the boundaries can be considered such that further integrations by parts incorporates the boundaries $\partial^2\mathcal{B}_0$ and $\partial^3\mathcal{B}_0$ which represent curves and points, see e.g. Schulte et al. [106] and Javili et al. [67]. Assuming that the principle of virtual work $\delta W^{\text{int}} - \delta W^{\text{ext}} = 0$ is valid with respect to the corresponding functional spaces of admissible solution and test functions, the external contribution can be formulated as

$$\delta W^{\text{ext}} = \int_{\mathcal{B}_0} \boldsymbol{B}\cdot\delta\boldsymbol{\varphi} + \int_{\Gamma_0^T} \bar{\boldsymbol{T}}\cdot\delta\boldsymbol{\varphi} \, \mathrm{d}A + \int_{\Gamma_0^M} \bar{\boldsymbol{M}} : \nabla\delta\boldsymbol{\varphi} \, \mathrm{d}A, \tag{5.131}$$

where $\boldsymbol{B}$ is a given body force per unit volume of the reference configuration. Moreover, $\bar{\boldsymbol{T}}$ and $\bar{\boldsymbol{M}}$ are a surface traction and a surface torque acting at the mechanical Neumann boundaries Γ_0^T and Γ_0^M. Eventually, we obtain the local form of the mechanical problem as

$$\nabla\cdot(\boldsymbol{\tau}\boldsymbol{F}^{-\mathrm{T}} - \nabla\cdot\mathfrak{P}) + \boldsymbol{B} = \boldsymbol{0} \tag{5.132}$$

1. Stress equilibrium
$$\nabla \cdot (\boldsymbol{\tau F}^{-\mathrm{T}} - \nabla \cdot \mathfrak{P}) + \boldsymbol{B} = \boldsymbol{0} \tag{5.136}$$
2. Kirchhoff stress
$$\boldsymbol{\tau} = \boldsymbol{\tau}_{\mathrm{mat}} + \boldsymbol{\tau}_{\mathrm{fib}}, \quad \boldsymbol{\tau}_{\mathrm{mat}} = \zeta \sum_a \lambda_a \frac{\partial \Psi_{\mathrm{mat}}}{\partial \lambda_a} \boldsymbol{n}_a \otimes \boldsymbol{n}_a, \quad \boldsymbol{\tau}_{\mathrm{fib}} = \frac{1-\zeta}{2} \frac{\partial \Psi_{\mathrm{fib}}}{\partial \boldsymbol{F}} \boldsymbol{F}^{\mathrm{T}} \tag{5.137}$$
3. Dirichlet and Neumann conditions
$$\begin{aligned} &\boldsymbol{\varphi} = \bar{\boldsymbol{\varphi}}(\boldsymbol{X}, t) \text{ on } \Gamma_0^{\varphi}, && (\boldsymbol{\tau F}^{-\mathrm{T}} - \nabla \cdot \mathfrak{P})\boldsymbol{N} = \bar{\boldsymbol{T}}(\boldsymbol{X}, t) \text{ on } \Gamma_0^T \\ &\nabla \boldsymbol{\varphi N} = \nabla \bar{\boldsymbol{\varphi}}(\boldsymbol{X}, t)\boldsymbol{N} \text{ on } \Gamma_0^{\nabla\varphi}, && \mathfrak{P}\boldsymbol{N} = \bar{\boldsymbol{M}}(\boldsymbol{X}, t) \text{ on } \Gamma_0^M \end{aligned} \tag{5.138}$$
4. Initial conditions
$$\boldsymbol{\varphi}(\boldsymbol{X}, 0) = \boldsymbol{\varphi}_0, \ \dot{\boldsymbol{\varphi}}(\boldsymbol{X}, 0) = \boldsymbol{v}_0, \tag{5.139}$$

Table 5.3: Strong formulation of the problem.

supplemented by boundary conditions

$$\begin{aligned} \boldsymbol{\varphi} &= \bar{\boldsymbol{\varphi}} \quad \text{on} \quad \Gamma_0^{\varphi} \\ \nabla \boldsymbol{\varphi N} &= \nabla \bar{\boldsymbol{\varphi}} \boldsymbol{N} \quad \text{on} \quad \Gamma_0^{\nabla\varphi} \\ (\boldsymbol{\tau F}^{-\mathrm{T}} - \nabla \cdot \mathfrak{P})\boldsymbol{N} &= \bar{\boldsymbol{T}} \quad \text{on} \quad \Gamma_0^T \\ \mathfrak{P}\boldsymbol{N} &= \bar{\boldsymbol{M}} \quad \text{on} \quad \Gamma_0^M \end{aligned} \tag{5.133}$$

with prescribed fields $\bar{\boldsymbol{\varphi}}$ and $\nabla \bar{\boldsymbol{\varphi}} \boldsymbol{N}$ at the mechanical Dirichlet boundaries Γ_0^{φ} and $\Gamma_0^{\nabla\varphi}$. As usual for fourth-order boundary value problems, the entire boundary is decomposed twice, i.e. $\Gamma_0 = \Gamma_0^{\varphi} \cup \Gamma_0^T$ with $\Gamma_0^{\varphi} \cap \Gamma_0^T = \emptyset$ and $\Gamma_0 = \Gamma_0^{\nabla\varphi} \cup \Gamma_0^M$ with $\Gamma_0^{\nabla\varphi} \cap \Gamma_0^M = \emptyset$. For details related to the enforcement of the gradient condition given in $(5.133)_2$ see e.g. Schuß et al. [108]. The problem in the strong form is summarized in 5.3.

Weak Form

Based on the derivations concerning the mechanical field within the previous section, the set of admissible test functions related to $\delta\boldsymbol{\varphi}$ and its space are defined as

$$\mathcal{V}^{\varphi} = \{\delta\boldsymbol{\varphi} \in H^2(\mathcal{B}_0) \,|\, \delta\boldsymbol{\varphi} = \boldsymbol{0} \,\text{on}\, \Gamma_0^{\varphi}, \, \nabla \delta\boldsymbol{\varphi N} = \boldsymbol{0} \,\text{on}\, \Gamma_0^{\nabla\varphi}\}, \tag{5.134}$$

included within the Sobolev functional space of square integrable functions and derivatives H^k with $k \geq 0$. Then, the weak form of the multifield problem reads

$$\int_{\mathcal{B}_0} \nabla_x \delta\boldsymbol{\varphi} : \boldsymbol{\tau} + \nabla\nabla\delta\boldsymbol{\varphi} :\cdot \mathfrak{P} - \delta\boldsymbol{\varphi} \cdot \boldsymbol{B} \,\mathrm{d}V - \int_{\Gamma_0^T} \delta\boldsymbol{\varphi} \cdot \bar{\boldsymbol{T}} \,\mathrm{d}A - \int_{\Gamma_0^M} \nabla\delta\boldsymbol{\varphi} : \bar{\boldsymbol{M}} \,\mathrm{d}A = 0. \tag{5.135}$$

Note that we neglect inertia terms within the mechanical balance equation, i.e. we consider only quasi static problems.

5.2.3 Numerical examples

The following examples are dedicated to the verification of the higher-order bending contributions of the fiber material. In particular, we investigate the in-plane bending behavior using a benchmark test from the previous section with the shell formulation, as well as the out-of-plane bending behavior using a 4-point bending test. Additional examples demonstrating the capabilities of the proposed formulation for endless fiber-reinforced polymers are explored in the subsequent section.

In-plane bending test

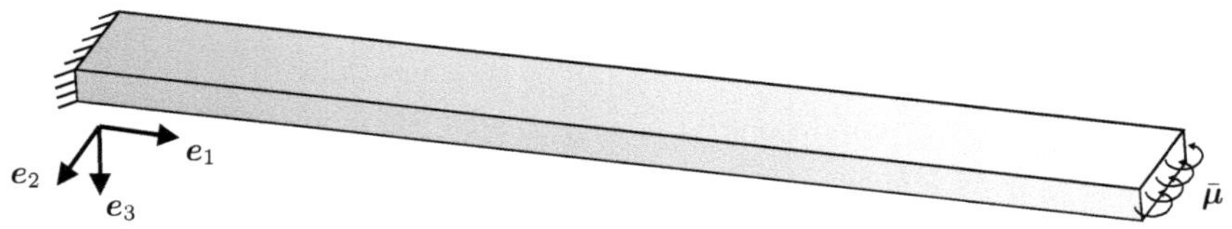

Figure 5.32: **In-plane bending test.** Problem setting. The lines illustrate the fiber structure.

We consider the same in-plane bending benchmark test as in Section 5.1.5, where the left edge of a Kirchhoff-Love shell is clamped while the right edge is subjected to an external in-plane torque, chosen to match a reference analytical solution. To verify the proposed formulation in terms of in-plane bending stiffness parameterization, we take the shell deformation result, extrude it to the corresponding 3D geometry and calculate the energy. The plate is of size $L \times W \times H = 10\,[\text{mm}] \times 1\,[\text{mm}] \times 0.5\,\text{mm}$ and is discretized by $8 \times 2 \times 1$ quadratic B-spline based elements, see Figure 5.32. Furthermore, we assume that the plate consists of a single fiber bundle with a cross section of $A = HW = 0.5\,\text{mm}^2$ and a tensile stiffness of $E_{\text{fib}} = 79000\,\text{N/mm}^2$. The area moments of inertia of the fiber bundle with respect to the $\boldsymbol{e}_3$-axis and the $\boldsymbol{e}_2$-axis are given by $I_{e_3} = HW^3/12 = 0.0417\,\text{mm}^4$ and $I_{e_2} = WH^3/12 = 0.0104\,\text{mm}^4$, respectively. Using these quantities we calibrate the bending stiffness parameters as $c_{\#} = E_{\text{fib}} I_{e_3}/A = 6583.3333\,\text{N}$ and $c_{\perp} = E_{\text{fib}} I_{e_2}/A = 2212\,\text{N}$.

In Figure 5.33, the strain energy density is depicted for both the Kirchhoff-Love shell formulation as well as the proposed higher-order continuum formulation. Therein, we can observe the same homogeneous distributions which verifies the calibration of the in-plane bending stiffness parameter $c_{\#}$. Note that the parameter $c_{\perp}$ does not contribute to the simulation result, but will be investigated within the next example.

4-point bending test

Next, the out-of-plane bending behavior of the fiber material is investigated using a 4-point bending test. Therefore, we consider again a rectangular geometry of size $L \times W \times$

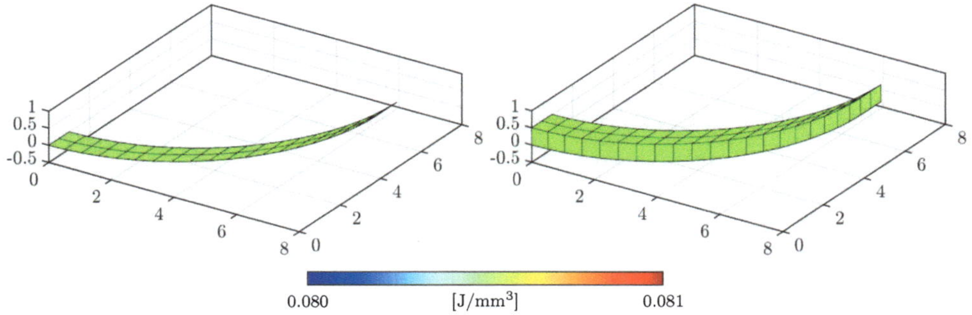

Figure 5.33: **In-plane bending test.** Strain energy distribution of the Kirchhoff-Love shell formulation (left) and higher-gradient continuum formulation (right).

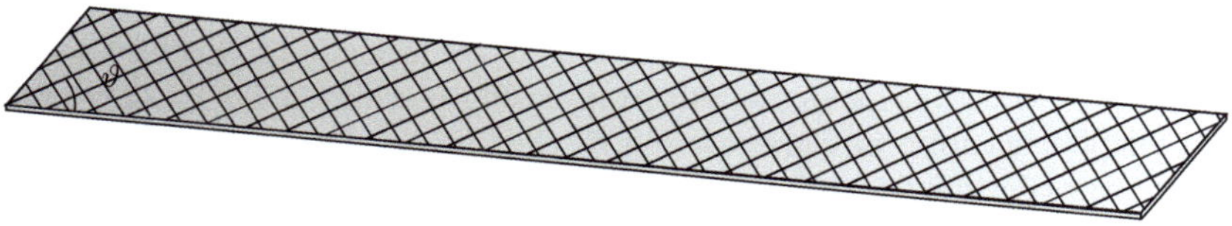

Figure 5.34: **In-plane bending test.** Problem setting. The lines illustrate the fiber structure.

$H = 125\,[\text{mm}] \times 25\,[\text{mm}] \times 0.5\,\text{mm}$ discretized by $50 \times 10 \times 2$ quadratic B-spline based elements. The bidirectional composite material has a matrix volume ratio of $\zeta = 0.53$ and the fibers are aligned in the $\vartheta = 0°$ configuration, see Figure 5.34 for the details on the fiber orientation. The four point bending test as shown in Figure 5.35 leads to a pure out-of-plane bending deformation of the structure. In particular, we prevent the displacement in upward direction for the outer support points and prescribe a displacement in downward direction for the inner contact points. Additionally, the left support point is horizontally fixed, whereas we allow sliding for the other contact points. The material setting of the matrix material reads $\mu = 1630.4\,\text{N/mm}^2$ and $\alpha_1 = 2$ for the deviatoric part and $\kappa = 6250\,\text{N/mm}^2$ and $\beta = -2$ for the volumetric part, which corresponds to a Young's modulus of $E_{\text{mat}} = 4500\,\text{N/mm}^2$ and a Poisson's ratio of $\nu = 0.38$.

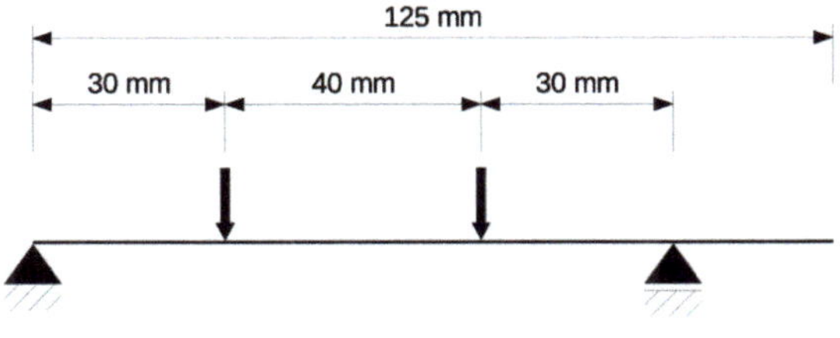

Figure 5.35: **4-point bending test.** Boundary conditions of the 4-point bending test.

Two different settings of the fiber material properties are applied assuming a single layer of fibers over thickness direction. Firstly, we set the tensile stiffness of the fibers to $a = E_{\text{fib}} = 79000\,\text{N/mm}^2$ and the bending stiffness to $c_\perp = 0\,\text{N}$. Secondly, we set the tensile stiffness of the fibers to zero and adjust the bending stiffness as $c_\perp = E_{\text{fib}} H^2/12 = 1645.83\,\text{N}$.

The applied bending stiffness of the continuum fiber model correlates to the out-of-plane bending stiffness for a shell model with the same height. As shown in the previous example, the proposed strain-gradient continuum formulation matches the contributions of a gradient shell formulation, provided that the stiffness is chosen properly. Thus, if we resolve the thickness of sufficiently flat geometry with in the continuum model to obtain the same deformation as expected for the shell theory, a coincident bending behavior of the structure should result. Figure 5.36 shows the load deflection result for the investigated material settings. As expected, both results match in a good agreement, i.e. the tension/compression behavior of the continuum fiber model in this bending example can be described by the bending terms themselves. This is an important and well known result, as strain-gradient contributions emanate from a length-scale dependent microstructure and if this microstructure is already resolved by the first-order continuum framework, the second-order contributions must be removed.

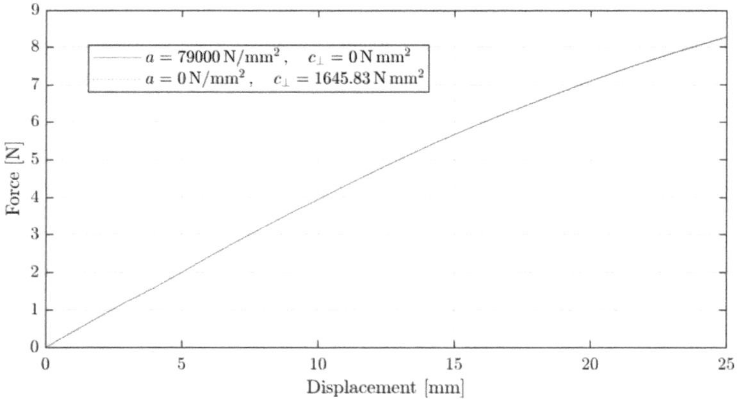

Figure 5.36: **4-point bending test.** Force-displacement curves for bending tests.

6 A Hybrid Phase-Field Model for Fracture in Fiber-Reinforced Polymers

Based on the concepts established in the preceding chapters, we now develop a comprehensive multifield framework for thermomechanical simulation of fiber-reinforced composites undergoing large deformations, and taking into account the microstructural size effects of the embeded fibers as well as inelastic, damage, and ultimate failure mechanisms. In particular, the polymeric matrix material is assumed to undergo porous-ductile fracture, while continuously embedded fibers undergo brittle fracture, as is characteristical e.g. for glass fiber-reinforced thermoplastics. A hybrid phase-field approach is developed and applied in conjunction with a modified Gurson-Tvergaard-Needelman GTN strain-gradient plasticity model that accounts for temperature-dependent growth of voids at the microscale. The mechanical response of the emerging woven fabric microstructure leads to additional higher-order terms representing homogenized bending contributions of the fibers. Considering the previous chapters, the underlying concepts and kinematic frameworks are only briefly introduced, with the focus instead on the multifield constitutive formulation for fiber-reinforced polymers. Eventually, a series of representative numerical examples, using a prototypical roving glass fiber-reinforced polyamide-6 (PA6) composite, is used to investigate the algorithmic performance and modeling capabilities, i.e. different kinds and sequences of failure within long fiber-reinforced polymers.

6.1 Configuration and kinematics

We consider a fiber-reinforced composite as a three-dimensional continuum body, along with the fields and kinematic framework introduced in Chapter 5 and Chapter 3. Assuming that the fiber deformation is congruent with the matrix deformation, we consider the joint field mapping $\boldsymbol{\varphi}(\boldsymbol{X}, t)$ and the material deformation gradient $\boldsymbol{F} = \nabla\boldsymbol{\varphi}(\boldsymbol{X}, t)$ with its determinant $J = \det(\boldsymbol{F}) > 0$. Moreover, we presume the absolute temperature $\theta(\boldsymbol{X}, t)$ as a further common field representing the thermal state of the matrix as well as the fiber material.

Regarding the damage behavior of the matrix material, the variables introduced in Chapter 3 describe the porous plasticity and ductile fracture of the matrix material. Namely, we consider the equivalent plastic strain $\alpha(\boldsymbol{X}, t)$ its dual, the dissipative resistance force

$r^{\mathrm{p}}(\boldsymbol{X},t)$, the plastic deformation map $\boldsymbol{F}^{\mathrm{p}}(\boldsymbol{X},t)$, the void volume fraction $f(\boldsymbol{X},t)$, and the crack phase-field $\mathfrak{s}(\boldsymbol{X},t)$.

Note that the biphasic character of semi-crystalline polymers, such as the matrix material polyamide-6 (PA6) considered in the numerical examples, is often modeled by a multi-mechanism (MM) constitutive model [20, 109] . Therein, the amorphous and crystalline phases are each characterized by their own inelastic strain field and porosity variable. Due to the significant influence of the fibers, we opted for a more straightforward approach for modeling the matrix material's mean mechanical effects, using a unified porous-elastoplastic model with averaged fields and material parameters, as in [73]. Furthermore, in agreement with the literature for the PA6 under study [68], porosity is assumed to be caused only by void growth. Consequently, void nucleation are neglected which simplifies the evolution laws of the isotropic hardening.

In addition, we propose a hybrid phase-field formulation for the fracture modeling of the fibers. Assuming the woven fiber reinforcement structure as in chapter ?, with the constant and orthogonal unit vector fields in the reference configuration of the two fiber filaments, $\boldsymbol{L}$ and $\boldsymbol{M}$, we introduce the order parameters

$$\mathfrak{s}_{\mathrm{L}}(\boldsymbol{X},t) : \mathcal{B}_0 \times \mathcal{T} \to \mathbb{R} \quad \text{with} \quad \mathfrak{s}_{\mathrm{L}} \in [0,1] \quad \text{and} \quad \dot{\mathfrak{s}}_{\mathrm{L}} \geq 0 \tag{6.1}$$

and

$$\mathfrak{s}_{\mathrm{M}}(\boldsymbol{X},t) : \mathcal{B}_0 \times \mathcal{T} \to \mathbb{R} \quad \text{with} \quad \mathfrak{s}_{\mathrm{M}} \in [0,1] \quad \text{and} \quad \dot{\mathfrak{s}}_{\mathrm{M}} \geq 0 \tag{6.2}$$

describing the crack phase-field of the fiber aligned in $\boldsymbol{L}$-direction and $\boldsymbol{M}$-direction. As for the matrix material, the values $\mathfrak{s} = 0$ refers to the undamaged and $\mathfrak{s} = 1$ to the fully ruptured state of the fiber material.

The variables introduced above characterize a multi-field setting for the formulation of temperature-dependent micro- and macromechanical damage in fiber-reinforced composites based on seven independent fields

$$\mathfrak{U} = [\boldsymbol{\varphi}, \theta, \alpha, r^{\mathrm{p}}, \mathfrak{s}, \mathfrak{s}_{\mathrm{L}}, \mathfrak{s}_{\mathrm{M}}], \tag{6.3}$$

the finite deformation map $\boldsymbol{\varphi}$, the absolute temperature field θ, the equivalent plastic strain field α, the dissipative plastic resistance force r^{p}, the crack phase-field $\mathfrak{s}$ of the matrix material, and the dual crack phase-field $[\mathfrak{s}_{\mathrm{L}}, \mathfrak{s}_{\mathrm{M}}]$ of the fiber material. The Lagrangian plastic deformation map $\boldsymbol{F}^{\mathrm{p}}$ and the void volume fraction f will be condensed within the balance equations.

The deformation measures for the porous-plastic fracture-degenerated matrix material are exactly the same as those introduced in Chapter 3. After a threefold multiplicative decomposition, namely an elasto-plastic decomposition, a volumetric-deviatoric decomposition and a fracture in-& sensitive decomposition over the eigenvalues, we finally obtain the elastic fracture insensitive part of the isochoric deformation gradient and the Jacobian determinant as

$$\tilde{\bar{\boldsymbol{F}}}^{\mathrm{e}} = \sum_a \tilde{\bar{\lambda}}^{\mathrm{e}}_a \boldsymbol{n}_a \otimes \boldsymbol{N}_a \quad \text{and} \quad \tilde{J}^{\mathrm{e}} = \begin{cases} \prod\limits_a (\lambda^{\mathrm{e}}_a)^{g(\mathfrak{s})} & \text{if} \quad \prod\limits_a \lambda^{\mathrm{e}}_a > 1 \\ \prod\limits_a \lambda^{\mathrm{e}}_a & \text{else} \end{cases}, \tag{6.4}$$

where $g = a_\mathrm{g}((1-\mathfrak{s})^3-(1-\mathfrak{s})^2)-2(1-\mathfrak{s})^3+3(1-\mathfrak{s})^2$ is the adjustable degradation function via the modeling parameter a_g. Moreover, $\tilde{\bar{\lambda}}^\mathrm{e}_a = (\bar{\lambda}^\mathrm{e}_a)^{g(\mathfrak{s})}$ are distortional stretches, defined via the isochoric, elastic parts of the principal stretches $\bar{\lambda}^\mathrm{e}_a = \prod_b(\lambda^\mathrm{e}_b)^{-1/3}\lambda^\mathrm{e}_a$, and lastly, the vectors $\boldsymbol{n}_a$ and $\boldsymbol{N}_a$ represent the principal directions of the left and right stretch tensors.

Regarding the fiber material, we use the deformation measures derived in Chapter 5. Along with the classical contributions for woven fiber structures considering the fiber directions, as in the stretches of the respective fiber threads

$$\lambda_\mathrm{L} = \|\boldsymbol{F}\boldsymbol{L}\| \quad \text{and} \quad \lambda_\mathrm{M} = \|\boldsymbol{F}\boldsymbol{M}\| \tag{6.5}$$

and the shear deformation

$$\phi = \tan\left(\operatorname{acos}\left(\frac{(\boldsymbol{F}\boldsymbol{L})\cdot(\boldsymbol{F}\boldsymbol{M})}{\|\boldsymbol{F}\boldsymbol{L}\|\|\boldsymbol{F}\boldsymbol{M}\|}\right)-\frac{\pi}{2}\right), \tag{6.6}$$

we additionally consider the curvature measures that take into account the deformed fiber vectors $\nabla\boldsymbol{l} = \nabla\boldsymbol{F}\boldsymbol{L}$ and $\nabla\boldsymbol{m} = \nabla\boldsymbol{F}\boldsymbol{M}$

$$\boldsymbol{\kappa}_\mathrm{L} = \frac{1}{\|\boldsymbol{F}\boldsymbol{L}\|^2}\left(\nabla\boldsymbol{F}\boldsymbol{L} - \frac{\boldsymbol{F}\boldsymbol{L}}{\|\boldsymbol{F}\boldsymbol{L}\|}\otimes\left(\frac{\boldsymbol{F}\boldsymbol{L}}{\|\boldsymbol{F}\boldsymbol{L}\|}\otimes\boldsymbol{L}\right):\nabla\boldsymbol{F}\right)\boldsymbol{L} \tag{6.7}$$

and

$$\boldsymbol{\kappa}_\mathrm{M} = \frac{1}{\|\boldsymbol{F}\boldsymbol{M}\|^2}\left(\nabla\boldsymbol{F}\boldsymbol{M} - \frac{\boldsymbol{F}\boldsymbol{M}}{\|\boldsymbol{F}\boldsymbol{M}\|}\otimes\left(\frac{\boldsymbol{F}\boldsymbol{M}}{\|\boldsymbol{F}\boldsymbol{M}\|}\otimes\boldsymbol{M}\right):\nabla\boldsymbol{F}\right)\boldsymbol{M}. \tag{6.8}$$

Next, we incorporate the hybrid crack phase-field into the fiber kinematics. Assuming that the fiber material is brittle compared to the matrix material and that fiber rupture requires a local tensile state, we formulate fracture insensitive parts of the stretches as

$$\tilde{\lambda}_\mathrm{L} = \begin{cases}(\lambda_\mathrm{L})^{g_\mathrm{L}(\mathfrak{s}_\mathrm{L})} & \text{if} \quad \lambda_\mathrm{L} > 1\\ \lambda_\mathrm{L} & \text{else}\end{cases} \quad \text{and} \quad \tilde{\lambda}_\mathrm{M} = \begin{cases}(\lambda_\mathrm{M})^{g_\mathrm{M}(\mathfrak{s}_\mathrm{M})} & \text{if} \quad \lambda_\mathrm{M} > 1\\ \lambda_\mathrm{M} & \text{else}\end{cases}, \tag{6.9}$$

where we use the same type of adjustable degradation function as for the matrix $g_\bullet = a_{\mathrm{g}_\bullet}((1-\mathfrak{s}_\bullet)^3-(1-\mathfrak{s}_\bullet)^2)-2(1-\mathfrak{s}_\bullet)^3+3(1-\mathfrak{s}_\bullet)^2$. Note that the $\bullet$ indicates the respective fiber direction $\boldsymbol{L}$ and $\boldsymbol{M}$. For the shear deformation we propose the following degradation

$$\tilde{\phi} = g_\mathrm{L}(\mathfrak{s}_\mathrm{L})g_\mathrm{M}(\mathfrak{s}_\mathrm{M})\phi, \tag{6.10}$$

where in case of single fiber rupture the shear is completely degraded even if the remaining fiber is undamaged. Finally, the fracture insensitive measures of the fiber curvatures take the straighforward form

$$\tilde{\boldsymbol{\kappa}}_\mathrm{L} = g_\mathrm{L}(\mathfrak{s}_\mathrm{L})\boldsymbol{\kappa}_\mathrm{L} \quad \text{and} \quad \tilde{\boldsymbol{\kappa}}_\mathrm{M} = g_\mathrm{M}(\mathfrak{s}_\mathrm{M})\boldsymbol{\kappa}_\mathrm{M}. \tag{6.11}$$

6.2 Variational formulation

Next, we propose the constitutive framework for thermomechanical damage in fiber-reinforced composites. To be specific, we introduce constitutive energetic and dissipative response functions based on the above definitions and derive the required relations and evolution laws to formulate the multifield variational problem.

6.2.1 Energetic response

The stored thermoelastic energy density of the composite material is defined by the functional

$$
\begin{aligned}
\Psi &= \zeta \Psi^{\mathrm{e},\theta}_{\mathrm{mat}}(\tilde{\bar{\boldsymbol{F}}}^{\mathrm{e}}, \tilde{J}^{\mathrm{e}}, \theta) + \frac{1-\zeta}{2} \Psi^{\mathrm{e},\theta}_{\mathrm{fib}}(\tilde{\lambda}_{\mathrm{L}}, \tilde{\lambda}_{\mathrm{M}}, \tilde{\phi}, \tilde{\boldsymbol{\kappa}}_{\mathrm{L}}, \tilde{\boldsymbol{\kappa}}_{\mathrm{M}}, \theta) \\
&= \zeta \left(\Psi^{\mathrm{e}}_{\mathrm{mat}}(\tilde{\bar{\boldsymbol{F}}}^{\mathrm{e}}, \tilde{J}^{\mathrm{e}}, \theta) + \Psi^{\theta}_{\mathrm{mat}}(\theta) \right) + \frac{1-\zeta}{2} \left(\Psi^{\mathrm{e}}_{\mathrm{fib}}(\tilde{\lambda}_{\mathrm{L}}, \tilde{\lambda}_{\mathrm{M}}, \tilde{\phi}, \tilde{\boldsymbol{\kappa}}_{\mathrm{L}}, \tilde{\boldsymbol{\kappa}}_{\mathrm{M}}, \theta) + \Psi^{\theta}_{\mathrm{fib}}(\theta) \right),
\end{aligned}
\tag{6.12}
$$

where $\zeta \in [0, 1]$ is the volume fraction of the matrix material. The elastic contribution to the stored energy function of the matrix material is decomposed into volumetric and deviatoric parts

$$
\Psi^{\mathrm{e}}_{\mathrm{mat}} = \Psi^{\mathrm{e,iso}}_{\mathrm{mat}} \left(\tilde{\bar{\boldsymbol{F}}}^{\mathrm{e}}(\lambda^{\mathrm{e}}_1, \lambda^{\mathrm{e}}_2, \lambda^{\mathrm{e}}_3, \mathfrak{s}), \theta \right) + \Psi^{\mathrm{e,vol}}_{\mathrm{mat}} \left(\tilde{J}^{\mathrm{e}}(\lambda^{\mathrm{e}}_1, \lambda^{\mathrm{e}}_2, \lambda^{\mathrm{e}}_3, \mathfrak{s}), \theta \right). \tag{6.13}
$$

As a representative non-linear constitutive law, a modified Ogden material model with the associated strain energy density function

$$
\Psi^{\mathrm{e,iso}}_{\mathrm{mat}} = \sum_a \sum_b \frac{\mu_b}{\alpha_b} \left((\tilde{\bar{\lambda}}^{\mathrm{e}}_a)^{\alpha_b} - 1 \right) \tag{6.14}
$$

and

$$
\Psi^{\mathrm{e,vol}}_{\mathrm{mat}} = \frac{\kappa}{\beta^2} \left(\beta \ln(\tilde{J}^{\mathrm{e}}) + (\tilde{J}^{\mathrm{e}})^{-\beta} - 1 \right) - 3 \frac{\epsilon \kappa}{\gamma} (\theta - \theta_0) \left((\tilde{J}^{\mathrm{e}})^{\gamma} - 1 \right) \tag{6.15}
$$

is used for the numerical examples. The parameters μ_b and α_b with $b = \{1, \ldots, N\}$ are related to the shear modulus and the parameters κ and β are related to the bulk modulus. Moreover, θ_0 is a reference temperature and the parameters ϵ and γ are related to the thermal expansion coefficient. Assuming that the fiber portion in both directions is identical, the corresponding elastic contribution of the fiber material is defined by

$$
\begin{aligned}
\Psi^{\mathrm{e}}_{\mathrm{fib}} &= \frac{1}{2} a \left((\tilde{\lambda}_{\mathrm{L}} - 1)^2 + (\tilde{\lambda}_{\mathrm{M}} - 1)^2 \right) + b \tilde{\phi}^2 \\
&+ \frac{1}{2} \left(\tilde{\boldsymbol{\kappa}}_{\mathrm{L}} \cdot \boldsymbol{c} \tilde{\boldsymbol{\kappa}}_{\mathrm{L}} + \tilde{\boldsymbol{\kappa}}_{\mathrm{M}} \cdot \boldsymbol{c} \tilde{\boldsymbol{\kappa}}_{\mathrm{M}} \right) + a v (\theta - \theta_0) \left((\tilde{\lambda}_{\mathrm{L}} - 1) + (\tilde{\lambda}_{\mathrm{M}} - 1) \right)
\end{aligned}
\tag{6.16}
$$

where v denotes the thermal expansion coefficient and a, b and $\boldsymbol{c}$ are stiffness parameters related to stretch, shear and curvature of the fiber material, as introduced in Chapter 5.

Next, the purely thermal contributions to the stored energy of the matrix and the fiber material are defined by

$$\Psi^{\theta}_{\mathrm{mat}} = c_{\mathrm{mat}} \left(\theta - \theta_0 - \theta \ln \left(\frac{\theta}{\theta_0} \right) \right) \tag{6.17}$$

and

$$\Psi^{\theta}_{\mathrm{fib}} = 2c_{\mathrm{fib}} \left(\theta - \theta_0 - \theta \ln \left(\frac{\theta}{\theta_0} \right) \right), \tag{6.18}$$

respectively. Therein, c_{mat} and c_{fib} are constant parameters representing specific heat capacities of the respective material.

The evolution of the stored thermoelastic energy is given by

$$\begin{aligned} \frac{\mathrm{d}}{\mathrm{d}t}\Psi = {} & \zeta \left(\sum_a \frac{\partial \Psi^{\mathrm{e}}_{\mathrm{mat}}}{\partial \lambda^{\mathrm{e}}_a} \dot{\lambda}^{\mathrm{e}}_a + \frac{\partial \Psi^{\mathrm{e}}_{\mathrm{mat}}}{\partial \mathfrak{s}} \dot{\mathfrak{s}} + \frac{\partial (\Psi^{\mathrm{e}}_{\mathrm{mat}} + \Psi^{\theta}_{\mathrm{mat}})}{\partial \theta} \dot{\theta} \right) \\ & + \frac{1-\zeta}{2} \left(\frac{\partial \Psi^{\mathrm{e}}_{\mathrm{fib}}}{\partial \boldsymbol{F}} \dot{\boldsymbol{F}} + \frac{\partial \Psi^{\mathrm{e}}_{\mathrm{fib}}}{\partial \nabla \boldsymbol{F}} \nabla \dot{\boldsymbol{F}} + \frac{\partial \Psi^{\mathrm{e}}_{\mathrm{fib}}}{\partial \mathfrak{s}_{\mathrm{L}}} \dot{\mathfrak{s}}_{\mathrm{L}} + \frac{\partial \Psi^{\mathrm{e}}_{\mathrm{fib}}}{\partial \mathfrak{s}_{\mathrm{M}}} \dot{\mathfrak{s}}_{\mathrm{M}} + \frac{\partial (\Psi^{\mathrm{e}}_{\mathrm{fib}} + \Psi^{\theta}_{\mathrm{fib}})}{\partial \theta} \dot{\theta} \right). \end{aligned} \tag{6.19}$$

Regarding the partial derivatives therein, we introduce first relations related to the Kirchhoff stress $\boldsymbol{\tau} = \boldsymbol{\tau}_{\mathrm{mat}} + \boldsymbol{\tau}_{\mathrm{fib}}$ as

$$\begin{aligned} \boldsymbol{\tau}_{\mathrm{mat}} &= \boldsymbol{\tau}^{\mathrm{dev}}_{\mathrm{mat}} + \boldsymbol{\tau}^{\mathrm{vol}}_{\mathrm{mat}} \\ &= \sum_a \left(\tau^{\mathrm{dev}}_{\mathrm{mat},a} + \tau^{\mathrm{vol}}_{\mathrm{mat},a} \right) \boldsymbol{n}_a \otimes \boldsymbol{n}_a \\ &= \zeta \sum_a \lambda^{\mathrm{e}}_a \left(\frac{\partial \Psi^{\mathrm{e,iso}}_{\mathrm{mat}}}{\partial \lambda^{\mathrm{e}}_a} + \frac{\partial \Psi^{\mathrm{e,vol}}_{\mathrm{mat}}}{\partial \lambda^{\mathrm{e}}_a} \right) \boldsymbol{n}_a \otimes \boldsymbol{n}_a \end{aligned} \tag{6.20}$$

and

$$\boldsymbol{\tau}_{\mathrm{fib}} = \frac{1-\zeta}{2} \left(\frac{\partial \Psi^{\mathrm{e}}_{\mathrm{fib}}}{\partial \tilde{\lambda}_{\mathrm{L}}} \frac{\partial \tilde{\lambda}_{\mathrm{L}}}{\partial \boldsymbol{F}} + \frac{\partial \Psi^{\mathrm{e}}_{\mathrm{fib}}}{\partial \tilde{\lambda}_{\mathrm{M}}} \frac{\partial \tilde{\lambda}_{\mathrm{M}}}{\partial \boldsymbol{F}} + \frac{\partial \Psi^{\mathrm{e}}_{\mathrm{fib}}}{\partial \tilde{\phi}} \frac{\partial \tilde{\phi}}{\partial \boldsymbol{F}} + \frac{\partial \Psi^{\mathrm{e}}_{\mathrm{fib}}}{\partial \tilde{\boldsymbol{\kappa}}_{\mathrm{L}}} \frac{\partial \tilde{\boldsymbol{\kappa}}_{\mathrm{L}}}{\partial \boldsymbol{F}} + \frac{\partial \Psi^{\mathrm{e}}_{\mathrm{fib}}}{\partial \tilde{\boldsymbol{\kappa}}_{\mathrm{M}}} \frac{\partial \tilde{\boldsymbol{\kappa}}_{\mathrm{M}}}{\partial \boldsymbol{F}} \right) \boldsymbol{F}^{\mathrm{T}}, \tag{6.21}$$

the higher-order stress of the fiber material as

$$\mathfrak{P}_{\mathrm{fib}} = \frac{1-\zeta}{2} \left(\frac{\partial \Psi^{\mathrm{e}}_{\mathrm{fib}}}{\partial \tilde{\boldsymbol{\kappa}}_{\mathrm{L}}} \frac{\partial \tilde{\boldsymbol{\kappa}}_{\mathrm{L}}}{\partial \nabla \boldsymbol{F}} + \frac{\partial \Psi^{\mathrm{e}}_{\mathrm{fib}}}{\partial \tilde{\boldsymbol{\kappa}}_{\mathrm{M}}} \frac{\partial \tilde{\boldsymbol{\kappa}}_{\mathrm{M}}}{\partial \nabla \boldsymbol{F}} \right), \tag{6.22}$$

the driving force of the respective crack phase-field as

$$\mathcal{H} = -\zeta \frac{\partial \Psi^{\mathrm{e}}_{\mathrm{mat}}}{\partial \mathfrak{s}}, \quad \mathcal{H}_{\mathrm{L}} = -\frac{1-\zeta}{2} \frac{\partial \Psi^{\mathrm{e}}_{\mathrm{fib}}}{\partial \mathfrak{s}_{\mathrm{L}}}, \quad \mathcal{H}_{\mathrm{M}} = -\frac{1-\zeta}{2} \frac{\partial \Psi^{\mathrm{e}}_{\mathrm{fib}}}{\partial \mathfrak{s}_{\mathrm{M}}} \tag{6.23}$$

and the specific entropy as

$$\begin{aligned} \eta &= \eta_{\mathrm{mat}} + \eta_{\mathrm{fib}} \\ &= -\zeta \frac{\partial (\Psi^{\mathrm{e}}_{\mathrm{mat}} + \Psi^{\theta}_{\mathrm{mat}})}{\partial \theta} - \frac{1-\zeta}{2} \frac{\partial (\Psi^{\mathrm{e}}_{\mathrm{fib}} + \Psi^{\theta}_{\mathrm{fib}})}{\partial \theta}. \end{aligned} \tag{6.24}$$

Moreover, we introduce a dissipation function

$$\mathcal{D}_{\text{int}} = \nu_{\text{p}_{\text{mat}}} \boldsymbol{\tau}_{\text{mat}} : \boldsymbol{d}^{\text{p}} + \nu_{\text{f}_{\text{mat}}} \mathcal{H}\dot{\mathfrak{s}} + \nu_{\text{f}_{\text{fib}}} (\mathcal{H}_{\text{L}}\dot{\mathfrak{s}}_{\text{L}} + \mathcal{H}_{\text{M}}\dot{\mathfrak{s}}_{\text{M}}), \tag{6.25}$$

to account for a transfer of dissipated energy due to plastification and fracture into the thermal field, where $\boldsymbol{d}^{\text{p}}$ denotes the Eulerian plastic rate of deformation tensor. The above relations are derived in a thermodynamically consistent manner by assuming that the dissipated energy is completely transfered into the thermal field, i.e. by setting $\nu_{\text{p}_{\text{mat}}} = \nu_{\text{f}_{\text{mat}}} = \nu_{\text{f}_{\text{fib}}} = 1$. Note, however, that the plastic dissipation factor $\nu_{\text{p}_{\text{mat}}}$ is typically chosen in the range of 85% to 95% in the context of thermoplasticity, see e.g. [111, 130, 75]. In addition, based on experimental observations it is reasonable to set fracture dissipation factors to $\nu_{\text{f}_{\text{mat}}} < 1$ and $\nu_{\text{f}_{\text{fib}}} < 1$, see the discussion related to an energy transfer into the thermal field in [39, 107] and the references therein.

To model the plastic and fracture mechanical response, we introduce an auxiliary functional as

$$\widehat{\Psi} = \zeta \left(\widehat{\Psi}^{\text{p}}_{\text{mat}}(\alpha, \nabla\alpha, \theta) + \widehat{\Psi}^{\text{f}}_{\text{mat}}(\mathfrak{s}, \nabla\mathfrak{s}, \alpha) \right) + \frac{1-\zeta}{2} \widehat{\Psi}^{\text{f}}_{\text{fib}}(\mathfrak{s}_{\text{L}}, \nabla\mathfrak{s}_{\text{L}}, \mathfrak{s}_{\text{M}}, \nabla\mathfrak{s}_{\text{M}}). \tag{6.26}$$

As introduced in Chapter 3, the plastic contribution $\widehat{\Psi}^{\text{p}}_{\text{mat}}$ describes the response of isotropic strain-gradient hardening related to the matrix material, with the equivalent plastic strain α its gradient $\nabla\alpha$ and the plastic length scale l_{p}

$$\widehat{\Psi}^{\text{p}}_{\text{mat}}(\alpha, \nabla\alpha, \theta) = \int_0^\alpha y(\bar{\alpha}, \theta)\, \mathrm{d}\bar{\alpha} + y_0(\theta) \frac{l_{\text{p}}^2}{2} \|\nabla\alpha\|^2. \tag{6.27}$$

The isotropic local strain hardening function $y(\alpha, \theta)$ developed specifically for the semi-crystalline polyamide-6 in the framework of a multi-mechanism model [20, 109] is adapted for a unified approach (monomechanism) as in [73]. Furthermore the function is extended to thermoplasticity following [111, 102, 28]. In particular, we obtain the saturation-type function

$$y(\alpha, \theta) = y_0(\theta) + y_1(\theta)\exp[\omega_{\text{p}1}\alpha] + y_2(\theta)(1 - \exp[-\omega_{\text{p}2}\alpha]), \tag{6.28}$$

with the three temperature-dependent material parameters $y_0 > 0$, $y_1 \geq 0$ and $y_2 \geq 0$ defined as

$$\begin{aligned} y_0(\theta) &= y_0(\theta_{\text{ref}})(1 - \omega_{\text{t}0}(\theta - \theta_{\text{ref}})), \\ y_1(\theta) &= y_1(\theta_{\text{ref}})(1 - \omega_{\text{t}1}(\theta - \theta_{\text{ref}})), \\ y_2(\theta) &= y_2(\theta_{\text{ref}})(1 - \omega_{\text{t}2}(\theta - \theta_{\text{ref}})). \end{aligned} \tag{6.29}$$

Therein, the initial yield stress $y_0 + y_1$ determines the threshold of the effective elastic response, $y_2(\theta)(1-\exp[-\omega_{\text{p}2}\alpha])$ describes an initial hardening stage. $y_1(\theta)\exp[\omega_{\text{p}1}\alpha]$ allows for the simulation of rheo-hardening, in which large stretches of fibrils leads to an abrupt increase of stress. Moreover, $\omega_{\text{p}1}$ and $\omega_{\text{p}2}$ are saturation parameters and $\omega_{\text{t}0}$, $\omega_{\text{t}1}$ and $\omega_{\text{t}2}$ are thermal hardening/softening parameters.

Next, we formulate fracture contributions for the matrix as well as the fiber material. As in the previous chapters, we approximate a sharp crack surface $\Gamma_\bullet$ by a regularized functional

$$\widehat{\Gamma}_\bullet(\mathfrak{s}_\bullet, \nabla\mathfrak{s}_\bullet) = \int_{\mathcal{B}_0} \widehat{\gamma}_\bullet(\mathfrak{s}_\bullet, \nabla\mathfrak{s}_\bullet)\,\mathrm{d}V \quad \text{with} \quad \widehat{\gamma}_\bullet(\mathfrak{s}_\bullet, \nabla\mathfrak{s}_\bullet) = \frac{1}{2l_{\mathrm{f}_\bullet}}(\mathfrak{s}_\bullet^2 + l_{\mathrm{f}_\bullet}^2\|\nabla\mathfrak{s}_\bullet\|^2), \tag{6.30}$$

based on a specific crack regularization profile $\widehat{\gamma}_\bullet$ defined per unit volume of the reference configuration and the fracture length scale $l_{\mathrm{f}_\bullet}$ which controls the regularization. Concerning ductile fracture of the matrix material, we require that $l_\mathrm{p} \geq l_\mathrm{f}$ such that the regularized crack zone lies inside of the plastic zone. Using the regularization given in (6.30), the approximated fracture energy of the composite material reads

$$W^\mathrm{f} \approx \int_{\mathcal{B}_0} \zeta g_\mathrm{c}(\alpha)\widehat{\gamma}(\mathfrak{s}, \nabla\mathfrak{s}) + \frac{1-\zeta}{2}\left(g_{\mathrm{c}_\mathrm{L}}\widehat{\gamma}_\mathrm{L}(\mathfrak{s}_\mathrm{L}, \nabla\mathfrak{s}_\mathrm{L}) + g_{\mathrm{c}_\mathrm{M}}\widehat{\gamma}_\mathrm{M}(\mathfrak{s}_\mathrm{M}, \nabla\mathfrak{s}_\mathrm{M})\right)\mathrm{d}V. \tag{6.31}$$

Here, $g_{\mathrm{c}_\bullet}$ denotes the Griffith-type critical energy density required to create fracture within the respective material. For the matrix material, the critical energy density is decomposed additively into elastic and plastic contributions as $g_\mathrm{c}(\alpha) = g_{\mathrm{c,p}} + g_{\mathrm{c,e}}\exp(-\omega_\mathrm{f}\alpha)$, with the modeling parameter ω_f. Summarized, the phase-field fracture contributions are given in terms of crack density functions as

$$\begin{aligned}\widehat{\Psi}^\mathrm{f}_\mathrm{mat} &= g_\mathrm{c}(\alpha)\widehat{\gamma}(\mathfrak{s}, \nabla\mathfrak{s}) \\ &= \frac{g_\mathrm{c}(\alpha)}{2l_\mathrm{f}}(\mathfrak{s}^2 + l_\mathrm{f}^2\|\nabla\mathfrak{s}\|^2)\end{aligned} \tag{6.32}$$

and

$$\begin{aligned}\widehat{\Psi}^\mathrm{f}_\mathrm{fib} &= g_{\mathrm{c}_\mathrm{L}}\widehat{\gamma}_\mathrm{L}(\mathfrak{s}_\mathrm{L}, \nabla\mathfrak{s}_\mathrm{L}) + g_{\mathrm{c}_\mathrm{M}}\widehat{\gamma}_\mathrm{M}(\mathfrak{s}_\mathrm{M}, \nabla\mathfrak{s}_\mathrm{M}) \\ &= \frac{g_{\mathrm{c}_\mathrm{L}}}{2l_{\mathrm{f}_\mathrm{L}}}(\mathfrak{s}_\mathrm{L}^2 + l_{\mathrm{f}_\mathrm{L}}^2\|\nabla\mathfrak{s}_\mathrm{L}\|^2) + \frac{g_{\mathrm{c}_\mathrm{M}}}{2l_{\mathrm{f}_\mathrm{M}}}(\mathfrak{s}_\mathrm{M}^2 + l_{\mathrm{f}_\mathrm{M}}^2\|\nabla\mathfrak{s}_\mathrm{M}\|^2).\end{aligned} \tag{6.33}$$

Eventually, we obtain the dissipative resistance forces of the plastic field and the respective crack phase-field via the variational derivatives of $\widehat{\Psi}$ with respect to α and $\mathfrak{s}_\bullet$ as

$$r^\mathrm{p} = \zeta\delta_\alpha\widehat{\Psi}^\mathrm{p}_\mathrm{mat} = \zeta(\partial_\alpha\widehat{\Psi}^\mathrm{p}_\mathrm{mat} - \nabla\cdot(\partial_{\nabla\alpha}\widehat{\Psi}^\mathrm{p}_\mathrm{mat})) \tag{6.34}$$

and

$$r^\mathrm{f}_\bullet = \delta_{\mathfrak{s}_\bullet}\widehat{\Psi} = \partial_{\mathfrak{s}_\bullet}\widehat{\Psi} - \nabla\cdot(\partial_{\nabla\mathfrak{s}_\bullet}\widehat{\Psi}). \tag{6.35}$$

6.2.2 Dissipative response

The composite's dissipative response and evolution equations, which are very similar to the structure developed in Chapter 3, are now adapted and extended for fiber fracture. Regarding the porous elastoplastic behavior of the polyamide matrix material, multi-mechanism models [20, 109, 68] are commonly used in which local information such as plastic strains, stresses, and damages are taken into account in each phase of semi-crystalline polymers via several GTN-type yield functions [57, 123, 96]. Since in the context of this work, we are only interested in the mean mechanical effects of the matrix material, a unified GTN-type yield function [73] is used, which implicitly defines averaged quantities, i.e. the effective scalar stress $\bar{\sigma} := \bar{\sigma}(\boldsymbol{\sigma}_{\text{mat}}, f)$ in terms of the Cauchy stress tensor $\boldsymbol{\sigma}_{\text{mat}} = \boldsymbol{\tau}_{\text{mat}}/J$ and the void volume fraction f

$$\Upsilon^{\text{G}}(\bar{\sigma}, \boldsymbol{\sigma}_{\text{mat}}, f) = \frac{\sigma_{\text{eq}}^2}{\bar{\sigma}^2} + 2q_1 f \cosh\left[\frac{3}{2}q_2\frac{p}{\bar{\sigma}}\right] - \left(1 + (q_1 f)^2\right) = 0. \tag{6.36}$$

Here, $\sigma_{\text{eq}} = \sqrt{3/2}\,\|\boldsymbol{\tau}_{\text{mat}}^{\text{dev}}/J\|$ denotes the von Mises equivalent stress, $p = \frac{1}{3}\text{tr}[\boldsymbol{\tau}_{\text{mat}}/J]$ the pressure and $q_{1/2}$ are fitting parameters. Note that for $q_1 = 0$ the influence of the pressure and the void volume fraction vanishes, i.e. $\bar{\sigma} = \sigma_{\text{eq}}$. With the effective stress $\bar{\sigma}$ and the dissipative resistance force r^{p} we define the plastic yield function as

$$\Phi^{\text{p}}\left(\bar{\sigma}(\boldsymbol{\sigma}_{\text{mat}}, f), r^{\text{p}}\right) = \bar{\sigma} - r^{\text{p}}. \tag{6.37}$$

As standard practice for polymide [68], we neglect other influences such as void nucleation or void softening due to shear, and obtain the evolution form of the void growth $\dot{f} = (1-f)\text{tr}[\boldsymbol{d}^{\text{p}}]$. Following [85], the current void volume fraction is given in terms of the plastic deformation as

$$f = 1 - \frac{1 - f_0}{J^{\text{p}}}. \tag{6.38}$$

A plastic Lagrange multiplier λ^{p} is introduced to enforce the Karush-Kuhn-Tucker conditions

$$\lambda^{\text{p}} \geq 0, \quad \Phi^{\text{p}} \leq 0, \quad \lambda^{\text{p}}\Phi^{\text{p}} = 0. \tag{6.39}$$

For the incorporation of the fracture mechanical behavior, we define crack threshold functions as

$$\Phi^{\text{f}}_{\bullet}(\mathcal{H}_{\bullet} - r^{\text{f}}_{\bullet}) = \mathcal{H}_{\bullet} - r^{\text{f}}_{\bullet} \tag{6.40}$$

where the energetic driving forces $\mathcal{H}_{\bullet}$ are bounded by crack resistance forces $r^{\text{f}}_{\bullet}$ dual to the crack phase-field variables $\mathfrak{s}_{\bullet}$. Similar to plasticity, we introduce fracture Lagrange multipliers $\lambda^{\text{f}}_{\bullet}$ to enforce the Karush-Kuhn-Tucker conditions of the respective crack phase-field

$$\lambda^{\text{f}}_{\bullet} \geq 0, \quad \Phi^{\text{f}}_{\bullet} \leq 0, \quad \lambda^{\text{f}}_{\bullet}\Phi^{\text{f}}_{\bullet} = 0. \tag{6.41}$$

Based on the concept of maximum dissipation and the set $\mathfrak{C} = [\boldsymbol{\sigma}_{\text{mat}}, r^{\text{p}}, \mathcal{H} - r^{\text{f}}, \mathcal{H}_{\text{L}} - r^{\text{f}}_{\text{L}}, \mathcal{H}_{\text{M}} - r^{\text{f}}_{\text{M}}, \lambda^{\text{p}}, \lambda^{\text{f}}, \lambda^{\text{f}}_{\text{L}}, \lambda^{\text{f}}_{\text{M}}]$, we define an extended dissipation potential and obtain a constrained optimization problem as where the Lagrange multipliers λ^{p} and $\lambda^{\text{f}}_{\bullet}$ control the non-smooth evolution of plasticity and fracture, respectively. Then, the associated plastic evolution equations follows as

$$\boldsymbol{d}^{\text{p}} = \lambda^{\text{p}}\frac{\partial\Phi^{\text{p}}}{\partial\boldsymbol{\sigma}_{\text{mat}}} \quad \text{and} \quad \dot{\alpha} = -\frac{\lambda^{\text{p}}}{1-f}\frac{\partial\Phi^{\text{p}}}{\partial r^{\text{p}}} \tag{6.42}$$

and the evolution equation of the respective crack phase-field as

$$\dot{\mathfrak{s}}_{\bullet} = \lambda^{\text{f}}_{\bullet}\frac{\partial\Phi^{\text{f}}_{\bullet}}{\partial(\mathcal{H}_{\bullet} - r^{\text{f}}_{\bullet})}. \tag{6.43}$$

A penalty regularization of the Lagrange multipliers can be utilized as follows*

$$\lambda^{\mathrm{p}} = \frac{1}{\eta_{\mathrm{p}}}\langle\Phi^{\mathrm{p}}\rangle^{\mathrm{n_p}} \geq 0, \quad \lambda^{\mathrm{f}} = \frac{1}{\eta_{\mathrm{f}}}\langle\Phi^{\mathrm{f}}\rangle \geq 0, \quad \lambda^{\mathrm{f}}_{\mathrm{L}} = \frac{1}{\eta_{\mathrm{f_L}}}\langle\Phi^{\mathrm{f}}_{\mathrm{L}}\rangle \geq 0 \quad \text{and} \quad \lambda^{\mathrm{f}}_{\mathrm{M}} = \frac{1}{\eta_{\mathrm{f_M}}}\langle\Phi^{\mathrm{f}}_{\mathrm{M}}\rangle \geq 0, \tag{6.44}$$

where $\eta_{\mathrm{p}}, \mathrm{n_p}$ and $\eta_{\mathrm{f}_\bullet}$ are material parameters which characterize the viscosity of plastification and crack propagation. Note that in the sense of continuum setting as defined in (6.12) and (6.26), the rates obtained in (6.42) and (6.43) are weighted by the respective volume fraction ζ and $(1-\zeta)/2$, respectively.

6.2.3 Heat conduction

Assuming that the absolute temperature $\theta(\boldsymbol{X}, t)$ as a common field representing the thermal state of the matrix as well as the fiber material, we introduce a relation for the Piola-Kirchhoff heat flux vector as

$$\boldsymbol{Q}(\boldsymbol{F}, \theta, \mathfrak{s}) = -\boldsymbol{K}(\boldsymbol{F}, \mathfrak{s})\nabla\theta \tag{6.45}$$

which is known as Duhamel's law of heat conduction. The thermal conductivity tensor is defined as

$$\boldsymbol{K} = \left(K(1-\mathfrak{s}) + K^{\mathrm{conv}}\mathfrak{s}\right)\boldsymbol{F}^{-1}\boldsymbol{F}^{-\mathrm{T}}. \tag{6.46}$$

In case of fracture, the conduction degenerates locally such that we achieve a pure convection problem and the heat transfer depends mainly on the crack opening width of the matrix material. Here, we formulate the conductivity tensor $\boldsymbol{K}$ in terms of the phase-field parameter $\mathfrak{s}$. Moreover, $K = \zeta K_{\mathrm{mat}} + (1-\zeta)K_{\mathrm{fib}}$ is a average conductivity parameter related to the composite material and K^{conv} is a convection parameter.

6.2.4 Coupled problem

Based on the derivations concerning the mechanical field of the fiber-reinforced material just introduced, the set of admissible test functions related to $\mathfrak{U}$ is given as

$$\delta\mathfrak{U} = [\delta\boldsymbol{\varphi}, \delta\theta, \delta\alpha, \delta r^{\mathrm{p}}, \delta\mathfrak{s}, \delta\mathfrak{s}_{\mathrm{L}}, \delta\mathfrak{s}_{\mathrm{M}}], \tag{6.47}$$

i.e. variations of the deformation, the absolute temperature, the equivalent plastic strain, the dual plastic resistance force, the crack phase-field of the matrix material and the variables of the dual crack phase-field of the fiber material, where their spaces are defined

* The Macaulay brackets are defined by $\langle x\rangle = (x + |x|)/2$.

as

$$
\begin{aligned}
\mathcal{V}^{\varphi} &= \{\delta\boldsymbol{\varphi} \in H^2(\mathcal{B}_0) \,|\, \delta\boldsymbol{\varphi} = \mathbf{0} \,\text{on}\, \Gamma_0^{\varphi},\, \nabla\delta\boldsymbol{\varphi}\boldsymbol{N} = \mathbf{0} \,\text{on}\, \Gamma_0^{\nabla\varphi}\},\\
\mathcal{V}^{\theta} &= \{\delta\theta \in H^1(\mathcal{B}_0) \,|\, \delta\theta = 0 \text{ on } \Gamma_0^{\theta}\},\\
\mathcal{V}^{\alpha} &= \{\delta\alpha \in H^1(\mathcal{B}_0) \,|\, \delta\alpha = 0 \text{ on } \Gamma_0^{\alpha}\},\\
\mathcal{V}^{r^{\mathrm{p}}} &= \{\delta r^{\mathrm{p}} \in \mathcal{L}^2(\mathcal{B}_0)\},\\
\mathcal{V}^{\mathfrak{s}} &= \{\delta\mathfrak{s} \in H^1(\mathcal{B}_0) \,|\, \delta\mathfrak{s} = 0 \text{ on } \bar{\Gamma}\},\\
\mathcal{V}^{\mathfrak{s}_{\mathrm{L}}} &= \{\delta\mathfrak{s}_{\mathrm{L}} \in H^1(\mathcal{B}_0) \,|\, \delta\mathfrak{s}_{\mathrm{L}} = 0 \text{ on } \bar{\Gamma}_{\mathrm{L}}\},\\
\mathcal{V}^{\mathfrak{s}_{\mathrm{M}}} &= \{\delta\mathfrak{s}_{\mathrm{M}} \in H^1(\mathcal{B}_0) \,|\, \delta\mathfrak{s}_{\mathrm{M}} = 0 \text{ on } \bar{\Gamma}_{\mathrm{M}}\}
\end{aligned}
\tag{6.48}
$$

included within the Sobolev functional space of square integrable functions and derivatives H^k with $k \geq 0$. Then, the weak form of the coupled multifield problem reads

$$
\begin{aligned}
&\int_{\mathcal{B}_0} \nabla_x\delta\boldsymbol{\varphi} : \boldsymbol{\tau} + \nabla\nabla\delta\boldsymbol{\varphi} :\cdot \boldsymbol{\mathfrak{P}} - \delta\boldsymbol{\varphi}\cdot\boldsymbol{B}\,\mathrm{d}V - \int_{\Gamma_0^T} \delta\boldsymbol{\varphi}\cdot\bar{\boldsymbol{T}}\,\mathrm{d}A - \int_{\Gamma_0^M} \nabla\delta\boldsymbol{\varphi} : \bar{\boldsymbol{M}}\,\mathrm{d}A = 0,\\
&\int_{\mathcal{B}_0} \delta\theta(\theta\dot{\eta} - \mathcal{D}_{\mathrm{int}} - \mathcal{R}) - \nabla\delta\theta\cdot\boldsymbol{Q}\,\mathrm{d}V - \int_{\Gamma_0^Q} \delta\theta\bar{Q}\,\mathrm{d}A = 0,\\
&\int_{\mathcal{B}_0} \left(\delta\alpha(\zeta y - r^{\mathrm{p}}) + \zeta y_0 l_{\mathrm{p}}^2 \nabla\delta\alpha\cdot\nabla\alpha\right)\mathrm{d}V = 0,\\
&\int_{\mathcal{B}_0} \delta r^{\mathrm{p}}\left(\eta_{\mathrm{p}}\dot{\alpha} - \frac{\chi_{\mathrm{p}}(\Phi^{\mathrm{p}})^{\mathrm{n_p}}}{1-f}\right)\mathrm{d}V = 0,\\
&\int_{\mathcal{B}_0} \delta\mathfrak{s}\eta_{\mathrm{f}}\dot{\mathfrak{s}} - \delta\mathfrak{s}\chi_{\mathrm{f}}\left(\mathcal{H} - \frac{\zeta g_{\mathrm{c}}}{l_{\mathrm{f}}}\mathfrak{s}\right) + \chi_{\mathrm{f}}\zeta g_{\mathrm{c}}\nabla\delta\mathfrak{s}\cdot\nabla\mathfrak{s}\,\mathrm{d}V = 0,\\
&\int_{\mathcal{B}_0} \delta\mathfrak{s}_{\mathrm{L}}\eta_{\mathrm{f_L}}\dot{\mathfrak{s}}_{\mathrm{L}} - \delta\mathfrak{s}_{\mathrm{L}}\chi_{\mathrm{f_L}}\left(\mathcal{H}_{\mathrm{L}} - \frac{(1-\zeta)g_{\mathrm{c_L}}}{2l_{\mathrm{f_L}}}\mathfrak{s}_{\mathrm{L}}\right) + \chi_{\mathrm{f_L}}\frac{1-\zeta}{2}g_{\mathrm{c_L}}\nabla\delta\mathfrak{s}_{\mathrm{L}}\cdot\nabla\mathfrak{s}_{\mathrm{L}}\,\mathrm{d}V = 0,\\
&\int_{\mathcal{B}_0} \delta\mathfrak{s}_{\mathrm{M}}\eta_{\mathrm{f_M}}\dot{\mathfrak{s}}_{\mathrm{M}} - \delta\mathfrak{s}_{\mathrm{M}}\chi_{\mathrm{f_M}}\left(\mathcal{H}_{\mathrm{M}} - \frac{(1-\zeta)g_{\mathrm{c_M}}}{2l_{\mathrm{f_M}}}\mathfrak{s}_{\mathrm{M}}\right) + \chi_{\mathrm{f_M}}\frac{1-\zeta}{2}g_{\mathrm{c_M}}\nabla\delta\mathfrak{s}_{\mathrm{M}}\cdot\nabla\mathfrak{s}_{\mathrm{M}}\,\mathrm{d}V = 0,
\end{aligned}
\tag{6.49}
$$

where $\mathcal{R}$ is a given heat supply per unit volume of the reference configuration and $\bar{Q}$ is a heat supply across the thermal Neumann boundary Γ_0^Q. For each other field, homogeneous Neumann boundary conditions are applied and appropriate thermal Dirichlet boundary conditions are formulated in terms of prescribed temperature $\bar{\theta}$, see Table 6.1. Note that we neglect inertia terms within the mechanical balance equation, i.e. we consider only quasi static problems. Additionally, internal conditions for the crack phase-field equations are given by

$$
\mathfrak{s}_{\bullet} = 1 \quad \text{on} \quad \bar{\Gamma}_{\bullet} \subset \widehat{\Gamma}_{\bullet},
\tag{6.50}
$$

ensuring that a fully broken state remains broken. The Karush-Kuhn Tucker conditions in (6.39) and (6.41), are evaluated by inserting local variables defined as

$$\chi_{\mathrm{p}} =: \begin{cases} 1 & \text{for} \quad \Phi^{\mathrm{p}} > 0 \\ 0 & \text{otherwise} \end{cases} \quad \text{and} \quad \chi_{\mathrm{f}_\bullet} =: \begin{cases} 1 & \text{for} \quad \Phi^{\mathrm{f}}_\bullet > 0 \\ 0 & \text{otherwise} \end{cases} . \tag{6.51}$$

Using local variables $\chi_{\mathrm{f}_\bullet}$ in (6.49), we demand $\dot{\mathfrak{s}}_\bullet \geq 0$ for thermodynamical consistency, avoiding a transfer of dissipated energy back into the mechanical field. This prevents healing effects, which may be taken into account as well. We can also set $\chi_{\mathrm{f}_\bullet} \equiv 1$ and restrict only the fully broken state, i.e. we allow for healing until the respective crack phase-field reaches the value one.

6.3 Spatial discretization

Due to the inclusion of curvature contributions in the fiber material, the mechanical part of the variational problem requires approximation functions, which are globally at least C^1-continuous to satisfy $\boldsymbol{\varphi}^{\mathrm{h}}, \delta\boldsymbol{\varphi}^{\mathrm{h}} \in \mathcal{H}^2(\mathcal{B}_0^{\mathrm{h}})$. Therefore we apply the isogeometric analysis approach introduced in Section 2.5.2. Inserting the corresponding approximations and their variations ($\boldsymbol{\varphi}^{\mathrm{h}}, \delta\boldsymbol{\varphi}^{\mathrm{h}}, \mathfrak{s}_\bullet^{\mathrm{h}}, \delta\mathfrak{s}_\bullet^{\mathrm{h}}, \theta^{\mathrm{h}}, \delta\theta^{\mathrm{h}}, \alpha^{\mathrm{h}}, \delta\alpha^{\mathrm{h}}, r^{\mathrm{p,h}}, \delta r^{\mathrm{p,h}}$) into (6.49) yields the semi-discrete set of coupled equations

$$\begin{aligned}
&\delta\boldsymbol{q}_A \cdot \left[\int_{\mathcal{B}_0} \boldsymbol{\tau}^{\mathrm{h}} \nabla_x R^A + \mathfrak{P}^{\mathrm{h}} : \nabla\nabla R^A \,\mathrm{d}V - \boldsymbol{F}^{\mathrm{ext},A} \right] = 0, \\
&\delta\theta_A \left[\int_{\mathcal{B}_0} \left(\dot{\eta}^{\mathrm{h}} R^A R^B \theta_B - R^A \mathcal{D}^{\mathrm{h}}_{\mathrm{int}} - \nabla R^A \boldsymbol{Q}^{\mathrm{h}} \right) \mathrm{d}V - Q^{\mathrm{ext},A} \right] = 0, \\
&\delta\alpha_i \left[\int_{\mathcal{B}_0} N^i (\zeta y^{\mathrm{h}} - N^j r_j^{\mathrm{p}}) \,\mathrm{d}V + K_\alpha^{ij} \alpha_j \right] = 0, \\
&\delta r_i^{\mathrm{p}} \left[M_{r^{\mathrm{p}}}^{ij} \dot{\alpha}_j - \int_{\mathcal{B}_0} \chi_{\mathrm{p}} N^i \frac{(\Phi^{\mathrm{p,h}})^{n_{\mathrm{p}}}}{J^{\mathrm{h}}(1 - f^{\mathrm{h}})} \,\mathrm{d}V \right] = 0, \\
&\delta\mathfrak{s}_{\bullet,A} \left[M_{\mathfrak{s}_\bullet}^{AB} \dot{\mathfrak{s}}_{\bullet,B} - \int_{\mathcal{B}_0} R^A \mathcal{H}_\bullet^{\mathrm{h}} \,\mathrm{d}V + K_{\mathfrak{s}_\bullet}^{AB} \mathfrak{s}_{\bullet,B} \right] = 0.
\end{aligned} \tag{6.66}$$

Therein, $\boldsymbol{\tau}^{\mathrm{h}}$, $\mathfrak{P}^{\mathrm{h}}$, η^{h} and $\mathcal{H}_\bullet^{\mathrm{h}}$ are semi-discrete versions of the Kirchhoff stress tensor, the higher-order stress tensor, the local entropy and the phase-field driving forces obtained via the partial derivatives of the semi-discrete stored energy density

$$\Psi^{\mathrm{h}} = \Psi^{\mathrm{h}}(\tilde{\bar{\boldsymbol{F}}}^{\mathrm{e,h}}, \tilde{J}^{\mathrm{e,h}}, \theta^{\mathrm{h}}, \tilde{\lambda}_{\mathrm{L}}^{\mathrm{h}}, \tilde{\lambda}_{\mathrm{M}}^{\mathrm{h}}, \tilde{\phi}^{\mathrm{h}}, \tilde{\boldsymbol{\kappa}}_{\mathrm{L}}^{\mathrm{h}}, \tilde{\boldsymbol{\kappa}}_{\mathrm{M}}^{\mathrm{h}}), \tag{6.67}$$

1. Stress equilibrium
$$\nabla \cdot (\boldsymbol{\tau} \boldsymbol{F}^{-\mathrm{T}} - \nabla \cdot \mathfrak{P}) + \boldsymbol{B} = \boldsymbol{0} \tag{6.52}$$

2. Kirchhoff stress
$$\boldsymbol{\tau} = \boldsymbol{\tau}_{\mathrm{mat}} + \boldsymbol{\tau}_{\mathrm{fib}}, \quad \boldsymbol{\tau}_{\mathrm{mat}} = \zeta \sum_a \lambda_a^{\mathrm{e}} \frac{\partial \Psi_{\mathrm{mat}}^{\mathrm{e}}}{\partial \lambda_a^{\mathrm{e}}} \boldsymbol{n}_a \otimes \boldsymbol{n}_a, \quad \boldsymbol{\tau}_{\mathrm{fib}} = \frac{1-\zeta}{2} \frac{\partial \Psi_{\mathrm{fib}}^{\mathrm{e}}}{\partial \boldsymbol{F}} \boldsymbol{F}^{\mathrm{T}} \tag{6.53}$$

3. Higher-order stress
$$\mathfrak{P} = \frac{1-\zeta}{2} \frac{\partial \Psi_{\mathrm{fib}}^{\mathrm{e}}}{\partial \nabla \boldsymbol{F}} \tag{6.54}$$

4. Energy balance
$$\theta \dot{\eta} + \nabla \cdot \boldsymbol{Q} - \mathcal{D}_{\mathrm{int}} - \mathcal{R} = 0 \tag{6.55}$$

5. Entropy
$$\eta = \eta_{\mathrm{mat}} + \eta_{\mathrm{fib}}, \quad \eta_{\mathrm{mat}} = -\zeta \frac{\partial \left(\Psi_{\mathrm{mat}}^{\mathrm{e}} + \Psi_{\mathrm{mat}}^{\theta} \right)}{\partial \theta}, \quad \eta_{\mathrm{fib}} = -\frac{1-\zeta}{2} \frac{\partial \left(\Psi_{\mathrm{fib}}^{\mathrm{e}} + \Psi_{\mathrm{fib}}^{\theta} \right)}{\partial \theta}$$

6. Dissipation
$$\mathcal{D}_{\mathrm{int}} = \nu_{\mathrm{p_{mat}}} \boldsymbol{\tau}_{\mathrm{mat}} : \boldsymbol{d}^{\mathrm{p}} + \nu_{\mathrm{f_{mat}}} \mathcal{H} \dot{\mathfrak{s}} + \nu_{\mathrm{f_{fib}}} (\mathcal{H}_{\mathrm{L}} \dot{\mathfrak{s}}_{\mathrm{L}} + \mathcal{H}_{\mathrm{M}} \dot{\mathfrak{s}}_{\mathrm{M}}) \tag{6.56}$$

7. Piola-Kirchhoff heat flux
$$\boldsymbol{Q} = -\boldsymbol{K} \nabla \theta, \quad \boldsymbol{K} = \left(K(1-\mathfrak{s}) + K^{\mathrm{conv}} \mathfrak{s} \right) \boldsymbol{F}^{-1} \boldsymbol{F}^{-\mathrm{T}} \tag{6.57}$$

8. Plastic strain
$$\boldsymbol{d}^{\mathrm{p}} - \lambda^{\mathrm{p}} \frac{\partial \Phi^{\mathrm{p}}}{\partial \boldsymbol{\sigma}_{\mathrm{mat}}} = \boldsymbol{0}, \quad \lambda^{\mathrm{p}} = \frac{1}{\eta_{\mathrm{p}}} \langle \Phi^{\mathrm{p}} \rangle^{n_{\mathrm{p}}}, \quad \boldsymbol{\sigma}_{\mathrm{mat}} = \boldsymbol{\tau}_{\mathrm{mat}} / J \tag{6.58}$$

9. Equivalent plastic strain
$$-\dot{\alpha} - \frac{\lambda^{\mathrm{p}}}{1-f} \frac{\partial \Phi^{\mathrm{p}}}{\partial r^{\mathrm{p}}} = 0 \tag{6.59}$$

10. Plastic resistance force
$$r^{\mathrm{p}} = \zeta \delta_{\alpha} \widehat{\Psi}_{\mathrm{mat}}^{\mathrm{p}} \tag{6.60}$$

11. Crack phase-field equations
$$\dot{\mathfrak{s}}_{\bullet} - \lambda_{\bullet}^{\mathrm{f}} \frac{\partial \Phi_{\bullet}^{\mathrm{f}}}{\partial (\mathcal{H}_{\bullet} - r_{\bullet}^{\mathrm{f}})} = 0, \quad \lambda_{\bullet}^{\mathrm{f}} = \frac{1}{\eta_{\mathrm{f}_{\bullet}}} \langle \Phi_{\bullet}^{\mathrm{f}} \rangle \tag{6.61}$$

12. Crack phase-field driving forces
$$\mathcal{H}_{\bullet} = -\frac{\partial \Psi^{\mathrm{e}}}{\partial \mathfrak{s}_{\bullet}} \tag{6.62}$$

13. Fracture resistance forces
$$r_{\bullet}^{\mathrm{f}} = \delta_{\mathfrak{s}_{\bullet}} \widehat{\Psi} \tag{6.63}$$

14. Dirichlet and Neumann conditions
$$\begin{aligned} \boldsymbol{\varphi} &= \bar{\boldsymbol{\varphi}}(\boldsymbol{X}, t) \text{ on } \Gamma_0^{\varphi}, & (\boldsymbol{\tau} \boldsymbol{F}^{-\mathrm{T}} - \nabla \cdot \mathfrak{P}) \boldsymbol{N} &= \bar{\boldsymbol{T}}(\boldsymbol{X}, t) \text{ on } \Gamma_0^{T} \\ \nabla \boldsymbol{\varphi} &= \nabla \bar{\boldsymbol{\varphi}}(\boldsymbol{X}, t) \text{ on } \Gamma_0^{\nabla \varphi}, & \mathfrak{P} \boldsymbol{N} &= \bar{\boldsymbol{M}}(\boldsymbol{X}, t) \text{ on } \Gamma_0^{M} \\ \theta &= \bar{\theta}(\boldsymbol{X}, t) \text{ on } \Gamma_0^{\theta}, & -\boldsymbol{Q} \cdot \boldsymbol{N} &= \bar{Q} \text{ on } \Gamma_0^{Q} \\ \alpha &= 0 \text{ on } \Gamma_0^{\alpha}, & \nabla \alpha \cdot \boldsymbol{N} &= 0 \text{ on } \Gamma_0^{\nabla \alpha} \\ \mathfrak{s}_{\bullet} &= 1 \text{ on } \bar{\Gamma}_{\bullet}, & \nabla \mathfrak{s}_{\bullet} \cdot \boldsymbol{N} &= 0 \text{ on } \Gamma_0 \end{aligned} \tag{6.64}$$

15. Initial conditions
$$\boldsymbol{\varphi}(\boldsymbol{X}, 0) = \boldsymbol{\varphi}_0, \quad \dot{\boldsymbol{\varphi}}(\boldsymbol{X}, 0) = \boldsymbol{v}_0, \quad \theta(\boldsymbol{X}, 0) = \theta_0, \quad \alpha(\boldsymbol{X}, 0) = 0, \quad r^p(\boldsymbol{X}, 0) = 0, \quad \mathfrak{s}_{\bullet}(\boldsymbol{X}, 0) = 0 \tag{6.65}$$

Table 6.1: Strong formulation of the coupled problem

cf. (6.12)-(6.24). $\mathcal{D}^{\text{h}}_{\text{int}}$ and $\boldsymbol{Q}^{\text{h}}$ are semi-discrete definitions of the dissipation density and heat flux, cf. (6.25), (6.45) and (6.46). Moreover, the semi-discrete external contributions in $(6.66)_1$ and $(6.66)_2$ are formulated as

$$\boldsymbol{F}^{\text{ext},A} = \int_{\mathcal{B}_0} R^A \boldsymbol{B} \,\text{d}V + \int_{\Gamma_0^T} R^A \bar{\boldsymbol{T}} \,\text{d}A + \int_{\Gamma_0^M} \bar{\boldsymbol{M}} \nabla R^A \,\text{d}A \tag{6.68}$$

and

$$Q^{\text{ext},A} = \int_{\mathcal{B}_0} R^A \mathcal{R} \,\text{d}V + \int_{\partial\mathcal{B}_0^{\theta_\text{n}}} R^A \bar{Q} \,\text{d}A. \tag{6.69}$$

The coefficients of the matrices in $(6.66)_3$ and $(6.66)_4$ take the form

$$K_\alpha^{ab} = \zeta y_0 l_\text{p}^2 \int_{\mathcal{B}_0} \nabla N^a \cdot \nabla N^b \,\text{d}V \quad \text{and} \quad M_{r\text{P}}^{ab} = \eta_\text{p} \int_{\mathcal{B}_0} N^a N^b \,\text{d}V, \tag{6.70}$$

whereas the matrices in $(6.66)_5$ are given by

$$\begin{aligned}
M_{\mathfrak{s}_\bullet}^{AB} &= \eta_{\text{f}_\bullet} \int_{\mathcal{B}_0} R^A R^B \,\text{d}V, \\
K_{\mathfrak{s}}^{AB} &= \frac{\zeta}{l_\text{f}} \int_{\mathcal{B}_0} g_\text{c}^\text{h} \chi_\text{f} \left(R^A R^B + l_\text{f}^2 \nabla R^A \cdot \nabla R^B\right) \text{d}V, \\
K_{\mathfrak{s}_\text{L}}^{AB} &= \frac{1-\zeta}{2 l_{\text{f}_\text{L}}} \int_{\mathcal{B}_0} g_{\text{cL}}^\text{h} \chi_{\text{f}_\text{L}} \left(R^A R^B + l_{\text{f}_\text{L}}^2 \nabla R^A \cdot \nabla R^B\right) \text{d}V, \\
K_{\mathfrak{s}_\text{M}}^{AB} &= \frac{1-\zeta}{2 l_{\text{f}_\text{M}}} \int_{\mathcal{B}_0} g_{\text{cM}}^\text{h} \chi_{\text{f}_\text{M}} \left(R^A R^B + l_{\text{f}_\text{M}}^2 \nabla R^A \cdot \nabla R^B\right) \text{d}V.
\end{aligned} \tag{6.71}$$

Eventually, the semi-discrete functions $\widehat{y}^\text{h}$, $\Phi^{\text{p,h}}$ and g_c^h denote the local hardening, the plastic yield and the critical fracture energy density, cf. (6.28) and (6.37).

6.4 Temporal discretization

Following the framework established in Section 3.4 ,the semi-discrete coupled problem (6.66) is discretized in time to obtain a set of non-linear algebraic equations to be solved via a Newton-Raphson method

$$
\begin{aligned}
&\delta \boldsymbol{q}_A \cdot \left[\int_{\mathcal{B}_0} \boldsymbol{\tau}^{\mathrm{h}}_{n+1} (\nabla_x R^A)_{n+1} + \mathfrak{P}^{\mathrm{h}}_{n+1} \nabla\nabla R^A \, \mathrm{d}V - \boldsymbol{F}^{\mathrm{ext},A}_{n+1} \right] = 0, \\
&\delta \theta_A \left[\int_{\mathcal{B}_0} \left(\frac{\eta^{\mathrm{h}}_{n+1} - \eta^{\mathrm{h}}_n}{\Delta t} R^A R^B \theta_{B,n+1} - R^A \mathcal{D}^{\mathrm{h}}_{\mathrm{int},n+1} - \nabla R^A \boldsymbol{Q}^{\mathrm{h}}_{n+1} \right) \mathrm{d}V - Q^{\mathrm{ext},A}_{n+1} \right] = 0, \\
&\delta \alpha_i \left[\int_{\mathcal{B}_0} N^i (\zeta y^{\mathrm{h}}_{n+1} - N^j r^{\mathrm{p}}_{j,n+1}) \, \mathrm{d}V + K^{ij}_{\alpha} \alpha_{j,n+1} \right] = 0, \\
&\delta r^{\mathrm{p}}_i \left[M^{ij}_{r^{\mathrm{p}}} \frac{\alpha_{j,n+1} - \alpha_{j,n}}{\Delta t} - \int_{\mathcal{B}_0} \chi_{\mathrm{p},n+1} N^i \frac{(\Phi^{\mathrm{p,h}}_{n+1})^{n_{\mathrm{p}}}}{J^{\mathrm{h}}_{n+1}(1 - f^{\mathrm{h}}_{n+1})} \, \mathrm{d}V \right] = 0, \\
&\delta \mathfrak{s}_{\bullet,A} \left[M^{AB}_{\mathfrak{s}_\bullet} \frac{\mathfrak{s}_{\bullet,B,n+1} - \mathfrak{s}_{\bullet,B,n}}{\Delta t} - \int_{\mathcal{B}_0} R^A \mathcal{H}^{\mathrm{h}}_{\bullet,n+1} \, \mathrm{d}V + K^{AB}_{\mathfrak{s}_\bullet,n+1} \mathfrak{s}_{\bullet,B,n+1} \right] = 0.
\end{aligned}
\tag{6.72}
$$

Therein, a full-discrete definition of the internal dissipation using small values for the plastic viscosity parameter η_{p}, is given by

$$
\begin{aligned}
\mathcal{D}^{\mathrm{h}}_{\mathrm{int},n+1} :=\; & \nu_{\mathrm{p_{mat}}} J^{\mathrm{h}}_{n+1}(1 - f^{\mathrm{h}}_{n+1})\, r^{\mathrm{p,h}}_{n+1} N^a \frac{\alpha_{a,n+1} - \alpha_{a,n}}{\Delta t} + \nu_{\mathrm{f_{mat}}} \mathcal{H}^{\mathrm{h}}_{n+1} R^A \frac{\mathfrak{s}_{A,n+1} - \mathfrak{s}_{A,n}}{\Delta t} \\
& + \nu_{\mathrm{f_{fib}}} \left(\mathcal{H}^{\mathrm{h}}_{\mathrm{L},n+1} R^A \frac{\mathfrak{s}_{\mathrm{L},A,n+1} - \mathfrak{s}_{\mathrm{L},A,n}}{\Delta t} + \mathcal{H}^{\mathrm{h}}_{\mathrm{M},n+1} R^A \frac{\mathfrak{s}_{\mathrm{M},A,n+1} - \mathfrak{s}_{\mathrm{M},A,n}}{\Delta t} \right)
\end{aligned}
\tag{6.73}
$$

Note that we apply a staggered scheme for the solution of the multi-field problem, i.e. the displacement field along with the plastic and hardening fields $\{\boldsymbol{q}_{A,n+1}, \alpha_{i,n+1}, r^{\mathrm{p}}_{i,n+1}, \boldsymbol{F}^{\mathrm{p,h}}_{n+1}\}$, the crack phase-fields $\mathfrak{s}_{\bullet,A,n+1}$ and the temperature field $\theta_{A,n+1}$ are solved successively.

For the time integration of the plastic evolution equations, the exponential integration scheme introduced in the context of the return-mapping algorithm in Section 3.4 is formulated in terms of the eigenvalues as

$$
\lambda^{\mathrm{e}}_{a,n+1} = \lambda^{\mathrm{e}}_{a,\mathrm{tr}} \exp\left[-\Delta t \lambda^{\mathrm{p}}_{n+1} n_{a,n+1}\right] \quad \text{with} \quad n_{a,n+1} = \frac{\partial \Phi^{\mathrm{p}}_{n+1}}{\partial \sigma_{\mathrm{mat},a,n+1}}, \tag{6.74}
$$

where $\sigma_{\mathrm{mat},a,n+1} = (\tau^{\mathrm{dev}}_{\mathrm{mat},a,n+1} + \tau^{\mathrm{vol}}_{\mathrm{mat},a,n+1})/J_{n+1}$. Note that in contrast to standard von Mises plasticity $\boldsymbol{n}_{n+1} \neq \boldsymbol{n}_{\mathrm{tr}}$ and $\|\boldsymbol{n}_{n+1}\| \neq 1$, i.e. the plastic correction has to be performed by the Lagrange multiplier $\lambda^{\mathrm{p}}_{n+1}$ as well as the components $n_{a,n+1}$ which can be obtained by solving the non-linear relations

$$
\widehat{\Phi}^{\mathrm{p}}_{n+1} - \eta_{\mathrm{p}} \lambda^{\mathrm{p}}_{n+1} = 0 \quad \text{and} \quad \frac{\partial \Phi^{\mathrm{p}}_{n+1}}{\partial \sigma_{\mathrm{mat},a,n+1}} - n_{a,n+1} = 0 \tag{6.75}
$$

via an internal Newton-Raphson iteration.

6.5 Numerical examples

In this section, we investigate the accuracy and performance of the proposed formulation for endless fiber-reinforced polymers. After verifying the higher-order contributions of the fiber material using two simple bending tests in the previous Chapter, we now focus on a series of tensile tests that demonstrate the ability of the proposed hybrid phase-field model to investigate different failure mechanisms for a roving glass composite material depending on the fiber configuration. This study is complemented by thermal investigations of the damage behavior of the model and its influence on the ultimate failure. Without loss of generality, we use an neo-Hook's model for the matrix material in all examples, i.e., we put $b = 1$ and $\alpha_1 = 2$ in (6.14).

In particular, we consider a flat specimen of size $L \times W \times H = 125\,[\text{mm}] \times 25\,[\text{mm}] \times 2\,[\text{mm}]$. Figure 6.1 and 6.4 show the geometry in the reference configuration along with the applied boundary conditions and fiber configurations. Dirichlet boundary conditions are applied to the outer regions of length 20 [mm]. More specifically, one flap is fixed and the other flap is moved at a displacement rate of 0.5 [mm]/[s] within a quasi-static simulation setting neglecting inertial effects. The computational mesh consists of 2432 square NURBS elements. The material setting, representing a prototypical roving glass fiber-reinforced polyamide-6 (PA6) composite, is summarized in table 6.2. We assume a square cross-section of fibers with $A_{\text{fib}} = 0.0025\,\text{mm}^2$ and obtain a bending stiffness of $c_{\perp} = c_{\#} = E_{\text{fib}} A_{\text{fib}}/12 = 16.46\,[\text{N}]$.

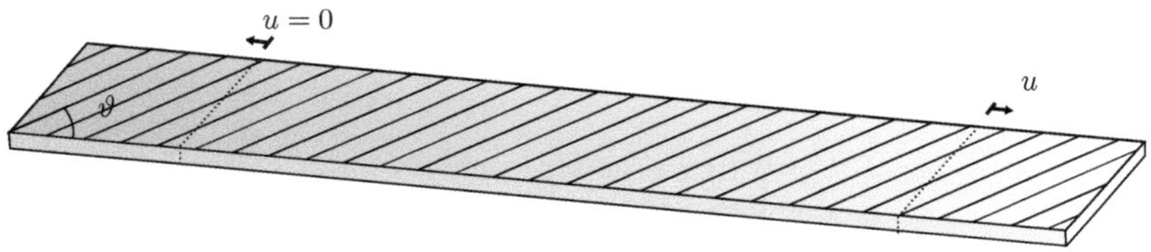

Figure 6.1: **Tensile Test (unidirectional).** Problem setting. The lines illustrate the fiber structure.

6.5.1 Unidirectional fiber reinforcement

We begin by examining the behavior of a unidirectionally reinforced composite with fiber orientations of $\vartheta = [0°, 10°, 20°, 30°, 40°, 65°, 90°]$, see Figure 6.1. The load deflection results for isothermal simulations at $\theta = 293\,\text{K}$ are shown in Figure 6.2. It denotes the crack initialization and final fracture of the fiber material by $\square$ and $\circ$, respectively. Also, crack initialization and final fracture of the matrix material are indicated by $\diamond$ and $\times$, respectively. Figure 6.3 shows the crack phase-field results of the fiber material and matrix material along with the plastic strain field results for the marked points. Note that the black mark indicates conditions without fiber fracture, since the specimen is already completely fractured.

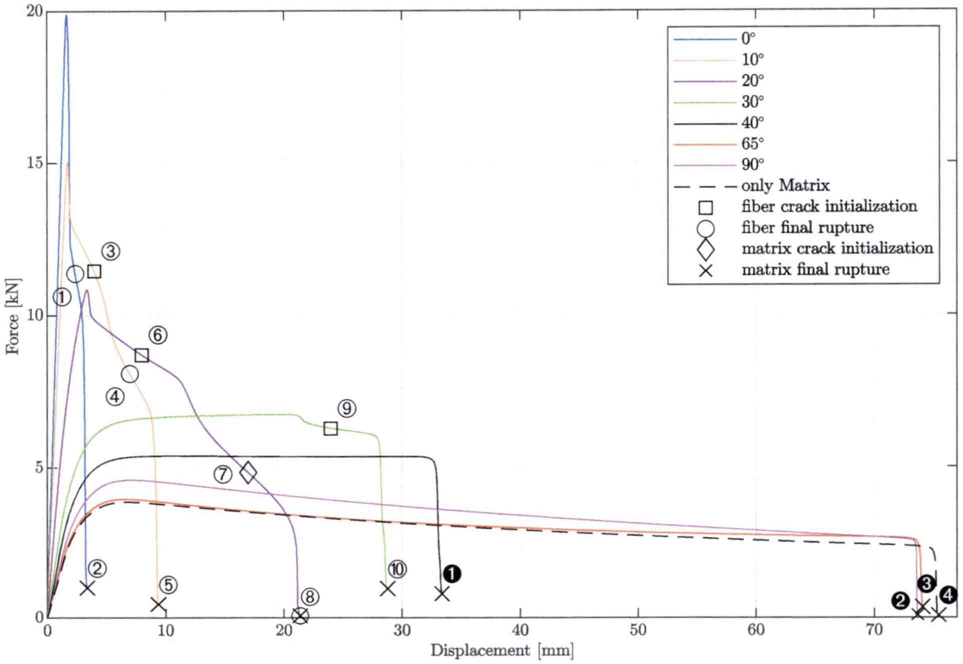

Figure 6.2: **Tensile Test (unidirectional).** Load deflection results for unidirectional fiber reinforcements with different orientations.

For a fiber orientation of $\vartheta = 0°$, the fibers account for most of the load transfer due to the different Young's modulus and fracture abruptly in the center of the specimen ①. Subsequently, the matrix material undergoes plastification and ductile fracture due to an abrupt load redistribution ②. Notably, the increased strain rates result in a strong viscoplastic behavior within the matrix material, which is regulated by the viscous regularization parameter.

In the unidirectional 10° fiber configuration, the fibers begin to crack near the clamping zones ③, which is further driven by the bending contribution to the crack driving force. This process is slowed down by the strain hardening behavior of the matrix material ④. In the ⑤ state, the fibers are fully ruptured and the matrix material starts to fracture in the same area.

At fiber orientation $\vartheta = 20°$, the brittle fracture of the fibers starts again near the clamping zones ⑥. However, more plastification and thus solidification of the matrix material ⑦ occurs, so that the fiber and matrix material undergo final rupture nearly at the same deformation state ⑧.

Applying a fiber orientation of $\vartheta = 30°$, direct load transfer between both boundaries by the fibers is not possible since fibers which are clamped at the lower end do not reach the

Figure 6.3: **Tensile Test (unidirectional).** Results of the fiber crack phase-field (first row), the plastic strain field (second row) and crack phase-field of the matrix material (third row). The results are shown for the different deformation states and fiber configurations marked in Figure 6.2.

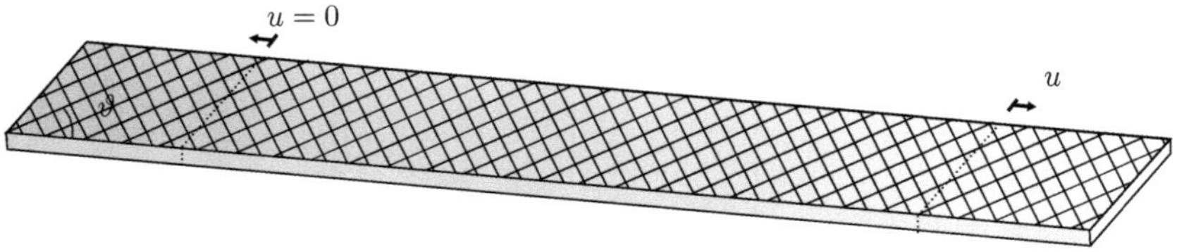

Figure 6.4: **Tensile Test (bidirectional).** Problem setting. The lines illustrate the fiber structure.

upper clamping zone. Hence, the load has to be transferred towards the matrix material, resulting in higher plastification ⑨ and ductile fracture at the center of the specimen ⑩. Note that the fibers begin to fracture only in small areas near the clamping zones ⑨.

For fiber orientations of $\vartheta = [40°, 65°, 90°]$, the fiber material does not fracture due to small loads acting in fiber direction ❶–❹. Instead, the matrix material undergoes suitable plastification and subsequently ductile fracture leading to failure orthogonal to the fiber orientations. Concerning the $\vartheta = 90°$ fiber configuration, the fibers controls the necking which can be observed by comparing the deformation with the results obtained for pure matrix material ❺. This is also evident in the marginally increased stiffness of the load deflection results prior to crack initiation, where the results are almost equal up to a displacement of $u = 40\,[\mathrm{mm}]$.

6.5.2 Bidirectional fiber reinforcement

Next, we examine the same tensile test using a bidirectionally reinforced material with fiber orientations of $\vartheta = [0°, 10°, 20°, 30°, 45°]$ as shown in Figure 6.4. The load- deformation results for isothermal simulations at $\theta = 293\,[\mathrm{K}]$ are shown in Figure 6.5. Again, the crack initialization and final fracture of the fiber material are represented by □ and ○, respectively. The final rupture of the matrix material is indicated by ×. The crack phase-field results of the fiber and matrix material as well as the plastic strain results are shown in Figure 6.6. Note that only the phase-field results of the fiber oriented in the ϑ direction are plotted.

For fiber orientations of $\vartheta = [0°, 10°, 20°]$, the bidirectionally reinforced material exhibits similar behavior to its corresponding unidirectionally reinforced counterparts. The additional orthogonal fiber merely provides higher necking resistance ①–⑧. In addition, the orthogonal fiber configuration restricts relative motion between the respective fibers, resulting in somewhat stiffer material behavior. This must be explored experimentally, which is outside the limits of the current work.

For fiber orientations of $\vartheta = [30°, 45°]$, this effect becomes more pronounced, as can be seen in Figure 6.5. As previously discussed, the additional orthogonal oriented fibers counteract the necking behavior due to the Poisson effect of the matrix material, causing

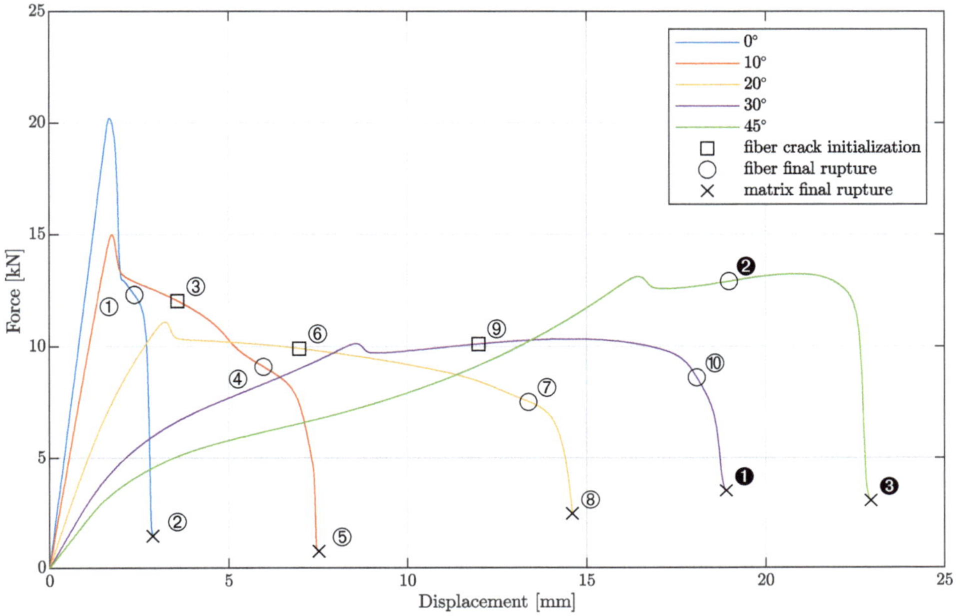

Figure 6.5: **Tensile Test (bidirectional).** Load deflection results for bidirectional, orthotropic fiber reinforcements with different orientations.

fractures within the matrix material near the clamping zones rather than in the center of the specimen ⑨–❸.

6.5.3 Thermal investigation

Finally, we analyse the temperature dependence of the proposed model. For this purpose, we repeat the tension test with a unidirectional fiber reinforcement as shown in Figure 6.1 using a fiber orientation of $\vartheta = 30°$.

Figure 6.7 depicts the load deflection result for isothermal simulations using temperatures of $\theta = [253, 273, 293]\,[\mathrm{K}]$. The corresponding crack phase-field results of the fiber and matrix material as well as the plastic strain results are shown for the final deformation step in Figure 6.8. As previously observed, the matrix material experiences significant plastic deformations for $\theta = 293\,[\mathrm{K}]$, followed by fiber fracture in small areas near the clamping zones, and finally the matrix material experiences ductile fracture in the center of the specimen. Reduced temperatures increase the yield stress of the matrix material, resulting in higher elastic energy and thus earlier, less ductile fracture of the matrix material. Note that in isothermal simulations with $\theta = [253, 273]\,[\mathrm{K}]$, the matrix material fails prior to the occurrence of any fiber cracks.

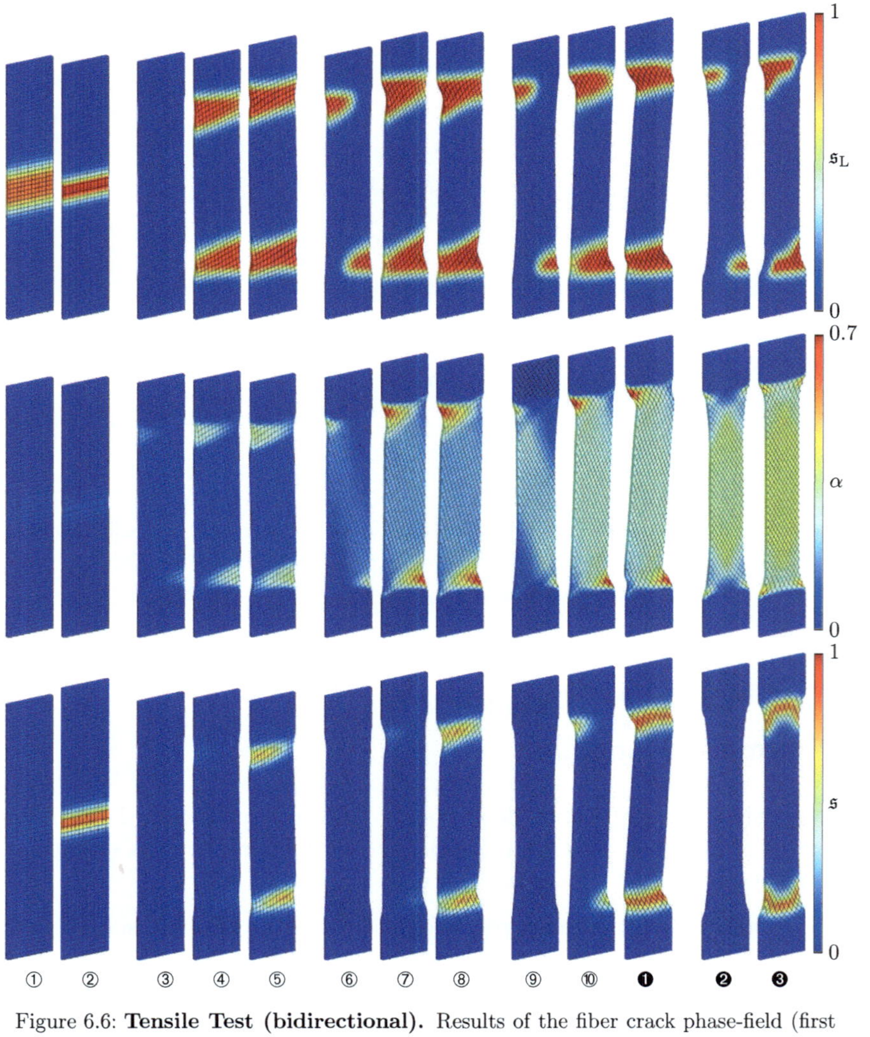

Figure 6.6: **Tensile Test (bidirectional).** Results of the fiber crack phase-field (first row) as well as the plastic strain field (second row) and crack phase-field of the matrix material (third row). The results are shown for the different deformation states and fiber configurations marked in Figure 6.5.

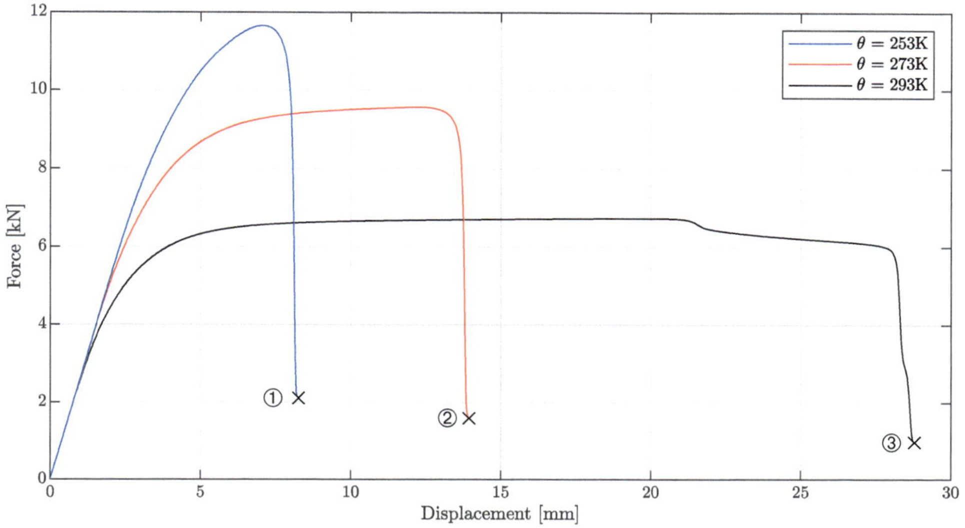

Figure 6.7: **Thermal investigations.** Load deflection results for isothermal simulations at different temperatures of $\theta = [253, 273, 293]$ [K] and a fiber orientation of $\vartheta = 30°$.

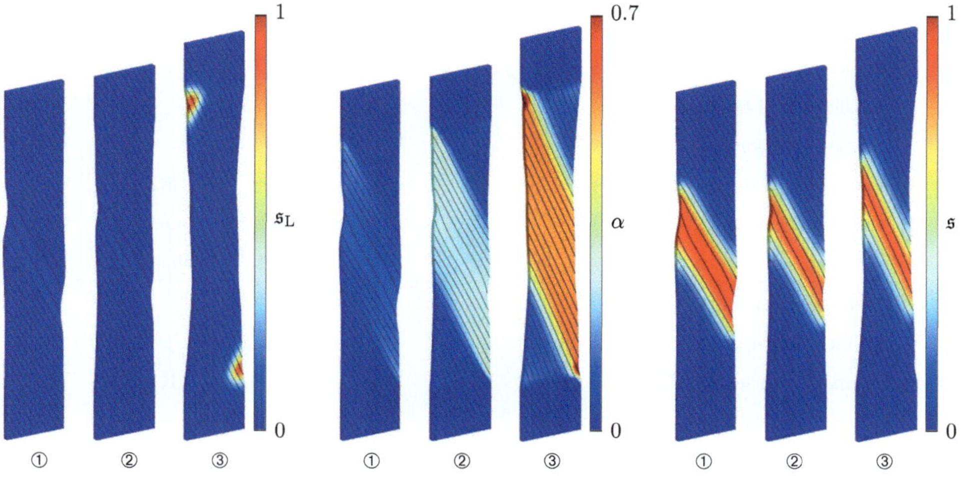

Figure 6.8: **Thermal investigations.** Results of isothermal simulations at different temperatures of $\theta = [253, 273, 293]$ [K] (each from left to right) and a fiber orientation of $\vartheta = 30°$. Results are shown for the fiber crack phase-field (first block), the plastic strain field (second block) and crack phase-field of the matrix material (third block) at the last deformation state marked in Figure 6.7.

Elastic parameters	
Shear modulus μ	1630 MPa
Shear exponent α	2
Bulk modulus κ	6250 MPa
Bulk parameter β	−2
Matrix volume ratio ζ	0.53
Tensile stiffness a	79000 MPa
Shear stiffness b	0 MPa
Bending stiffness $[c_\perp, c_\#]$	[16.46, 16.46] N
Plastic parameters	
Yield stress $[y_0(\theta_{\text{ref}}), y_1(\theta_{\text{ref}}), y_2(\theta_{\text{ref}})]$	[22, 56.8, 30] MPa
Saturation exponent $[\omega_{\text{p1}}, \omega_{\text{p2}}]$	[1, 115]
Thermal softening parameter $[\omega_{\text{t0}}, \omega_{\text{t1}}, \omega_{\text{t2}}]$	$[0.4, 0.4, 0.4]\,\text{K}^{-1}$
Viscoplastic parameter η_{p}	5000 MPa · s
Viscoplastic exponent n_{p}	1
Plastic length scale l_{p}	3.1 mm
Initial void fraction f_0	0.01
Gurson fitting parameter $[q_1, q_2]$	[3, 0.8]
Phase-field fracture parameters	
Brittle critical fracture energy $[g_{\text{c,e}}, g_{\text{c}_\text{L}}, g_{\text{c}_\text{M}}]$	$[500, 500, 500]\,\text{kJ/m}^2$
Ductile critical fracture energy $g_{\text{c,p}}$	$50\,\text{kJ/m}^2$
Saturation exponent ω_{f}	3
Fracture viscosity $[\eta_{\text{f}}, \eta_{\text{f}_\text{L}}, \eta_{\text{f}_\text{M}}]$	$[1, 1, 1] \cdot 10^{-7}\,\text{MPa} \cdot \text{s}$
Fracture length scale $[l_{\text{f}}, l_{\text{f}_\text{L}}, l_{\text{f}_\text{M}}]$	[3.1, 3.1, 3.1] mm
Degradation parameter $[a_{\text{g}}, a_{\text{g}_\text{L}}, a_{\text{g}_\text{M}}]$	[0.001, 0.001, 0.001]
Thermal parameters	
Specific heat capacity $[c_{\text{mat}}, c_{\text{fib}}]$	$[1860, 2080]\,\text{kJ/(m}^3 \cdot \text{K)}$
Thermal expansion coefficient $[\epsilon, \upsilon]$	$[106, 5] \cdot 10^{-6}\,\text{K}^{-1}$
Thermal expansion parameter γ	1
Conductivity K	0.25 W/(m · K)
Convection K_{conv}	0 W/(m · K)
Reference temperature θ_{ref}	293 K
Fracture & plastic dissipation factor $[\nu_{\text{p}_\text{mat}}, \nu_{\text{f}_\text{mat}}, \nu_{\text{f}_\text{fib}}]$	[0.9, 0.9, 0.9]

Table 6.2: Material setting of the fiber-reinforced composite (PA 6/Roving glass).

7 Summary and outlook

7.1 Summary

The non-linear framework presented in this work allows for a comprehensive investigation of damage and fracture in fiber-reinforced polymers. The combination of a second-gradient theory, a novel hybrid phase-field model and a temperature dependent GTN-type plasticity model provides a numerical framework which is able to describe different failure mechanisms in detail. This approach allows for improvements in the design of such composite materials since we are able to predict fiber and matrix failure and their sequence dependent on the fiber orientation. Moreover, due to the fully-coupled, thermomechanical microstructural approach we can optimize the fiber orientation for specific loads and thermal states. The verification and adjustment of the corresponding material parameters allows us to identify the physical effects of the higher-order terms in this prototypical second-gradient material. Several numerical tests conducted within this work have demonstrate the capability of the proposed framework to investigate such a complex behavior including the growth of microvoids, plasification and necking, crack initiation and propagation within the composite material and its components, respectively.

7.2 Outlook

Based on the present contribution the following research projects are currently in progress or in application:

- A framework for simulating thermoplastics with short fiber-reinforcements, which are easier and cheaper to produce than continuous fiber composites, is currently under development
- A variationally consistent multidimensional coupling approach for fiber-reinforced polymers where the Youngs moduli vary widely is under investigation
- Contact mechanics for second- and third-gradient materials based on the concept of mortar-type methods are in progress
- A research project on the simulation of the cracking process in bar-reinforced ultra-high performance concrete (UHPFRC) is currently in application

Bibliography

[1] E. C. Aifantis. On the Microstructural Origin of Certain Inelastic Models. *Journal of Engineering Materials and Technology*, 106(4):326–330, 1984.

[2] F. Aldakheel. *Mechanics of Nonlocal Dissipative Solids: Gradient Plasticity and Phase Field Modeling of Ductile Fracture.* PhD thesis, University of Stuttgart, 2016.

[3] F. Aldakheel. Micromorphic approach for gradient-extended thermo-elastic-plastic solids in the logarithmic strain space. *Continuum Mechanics and Thermodynamics*, 29(6):1207–1217, 2017.

[4] F. Aldakheel, S. Mauthe, and C. Miehe. Towards phase field modeling of ductile fracture in gradient-extended elastic-plastic solids. *Proceedings in Applied Mathematics and Mechanics*, 14:411–412, 2014.

[5] F. Aldakheel and C. Miehe. Coupled thermomechanical response of gradient plasticity. *International Journal of Plasticity*, 91:1–24, 2017.

[6] F. Aldakheel, P. Wriggers, and C. Miehe. A modified gurson-type plasticity model at finite strains: Formulation, numerical analysis and phase-field coupling. *Computational Mechanics*, 62:815–833, 2018.

[7] M. Ambati, R. Kruse, and L. De Lorenzis. A phase-field model for ductile fracture at finite strains and its experimental verification. *Computational Mechanics*, 57(1):149–167, Jan 2016.

[8] H. Amor, J.J. Marigo, and C. Maurini. Regularized formulation of the variational brittle fracture with unilateral contact: Numerical experiments. *Journal of the Mechanics and Physics of Solids*, 57:1209–1229, 2009.

[9] L. Anand, M. E. Gurtin, S. P. Lele, and C. Gething. A one-dimensional theory of strain-gradient plasticity: formulation, analysis, numerical results. *Journal of the Mechanics and Physics of Solids*, 53:1789–1826, 2005.

[10] T. Asmanoglo and A. Menzel. A finite deformation continuum modelling framework for curvature effects in fibre-reinforced nanocomposites. *Journal of the Mechanics and Physics of Solids*, 107:411–432, 2017.

[11] T. Asmanoglo and A. Menzel. A multi-field finite element approach for the modelling of fibre-reinforced composites with fibre-bending stiffness. *Computer Methods in Applied Mechanics and Engineering*, 317:1037–1067, 2017.

[12] F. Auricchio, L. Beirão da Veiga, J. Kiendl, C. Lovadina, and A. Reali. Locking-free isogeometric collocation methods for spatial Timoshenko rods. *Computer Methods in Applied Mechanics and Engineering*, 263:113–126, 2013.

[13] Z.P. Bazant and G. Pijaudier-Cabot. Measurement of Characteristic Length of Nonlocal Continuum. *Journal of Engineering Mechanics*, 115:755–767, 1989.

[14] K.U. Bletzinger, M. Bischoff, and E. Ramm. A unified approach for shear-locking-free triangular and rectangular shell finite elements. *Computers & Structures*, 75(3):321–334, 2000.

[15] C.G. Böhmer, Y. Lee, and P. Neff. Soliton solutions in geometrically nonlinear Cosserat micropolar elasticity with large deformations. *Wave motion*, 84:110–124, 2019.

[16] J. Bonet, A.J. Gil, and R. Ortigosa. A computational framework for polyconvex large strain elasticity. *Computer Methods in Applied Mechanics and Engineering*, 283:1061–1094, 2015.

[17] J. Bonet and R. D. Wood. *Nonlinear continuum mechanics for finite element analysis*. Cambridge University Press, second edition, 2008.

[18] M.J. Borden. *Isogeometric Analysis of Phase-Field Models for Dynamic Brittle and Ductile Fracture*. PhD thesis, The University of Texas at Austin, 2012.

[19] M.J. Borden, T.J.R. Hughes, C.M. Landis, and C.V. Verhoosel. A higher-order phase-field model for brittle fracture: Formulation and analysis within the isogeometric analysis framework. *Computer Methods in Applied Mechanics and Engineering*, 273:100–118, 2014.

[20] H.-A. Cayzac, K. Saï, and L. Laiarinandrasana. Damage based constitutive relationships in semi-crystalline polymer by using multi-mechanisms model. *International Journal of Plasticity*, 51:47–64, 2013.

[21] A. Cemal Eringen. *Microcontinuum Field Theories*. Springer-Verlag New York, 1999.

[22] J. Choo and W Sun. Coupled phase-field and plasticity modeling of geological materials: From brittle fracture to ductile flow. *Computer Methods in Applied Mechanics and Engineering*, 330:1–32, 2018.

[23] P. G. Ciarlet. *Mathematic Elasticity Volume I*. Noth-Holland, Amsterdam, 1988.

[24] P. G. Ciarlet. *An Introduction to Differential Geometry with Aplications to Elasticity*. Springer, 2005.

[25] E. Cosserat and F. Cosserat. *Théorie de Corps déformables*. Librairie Scientifique A. Hermann et fils, 1909.

[26] J.A. Cottrell, T.J.R. Hughes, and Y. Bazilevs. *Isogeometric Analysis: Toward Integration of CAD and FEA*. Wiley, New York, 2009.

[27] M. G. Cox. The numerical evaluation of B-splines. *IMA Journal of Applied Mathematics*, 10(2):134–149, 1972.

[28] H.S. Da Costa Mattos, J.F. Brandao, F. Amorim, P.V. De Souza Araujo, and J. Reis. A unified expression to estimate the stress-strain curve of polyamides at different temperatures. *Materials Research Express*, 6, 2019.

[29] R. de Boer. *Vektor- und Tensorrechnung für Ingenieure*. Springer-Verlag Berlin, Heidelberg, 1982.

[30] R. de Borst and H.B. Mühlhaus. Gradient-dependent plasticity: formulation and algorithmic aspects. *International Journal for Numerical Methods in Engineering*, 35:521–539, 1992.

[31] F. dell'Isola, I. Giorgio, M. Pawlikowski, and N.L. Rizzi. Large deformations of planar extensible beams and pantographic lattices: heuristic homogenization, experimental and numerical examples of equilibrium. *Proceedings of the Royal Society of London A: Mathematical, Physical and Engineering Sciences*, 472(2185), 2016.

[32] F. dell'Isola, P. Seppecher, M. Spagnuolo, E. Barchiesi, F. Hild, T. Lekszycki, I. Giorgio, L. Placidi, U. Andreaus, M. Cuomo, S.R. Eugster, A. Pfaff, K. Hoschke, R. Langkemper, E. Turco, R. Sarikaya, A. Misra, M. De Angelo, F. D'Annibale, A. Bouterf, X. Pinelli, A. Misra, B. Desmorat, M. Pawlikowski, C. Dupuy, D. Scerrato, P. Peyre, M. Laudato, L. Manzari, P. Göransson, C. Hesch, S. Hesch, P. Franciosi, J. Dirrenberger, F. Maurin, Z. Vangelatos, C. Grigoropoulos, V. Melissinaki, M. Farsari, W. Muller, B.E. Abali, C. Liebold, G. Ganzosch, P. Harrison, R. Drobnicki, L. Igumnov, F. Alzahrani, and T. Hayat. Advances in pantographic structures: design, manufacturing, models, experiments and image analyses. *Continuum Mechanics and Thermodynamics*, 31:1231–1282, 2019.

[33] M. Diehl, M. Wicke, P. Shanthraj, F. Roters, A. Brueckner-Foit, and D. Raabe. Coupled crystal plasticity–phase field fracture simulation study on damage evolution around a void: Pore shape versus crystallographic orientation. *JOM*, 69(5):872–878, 2017.

[34] B. J. Dimitrijevic and K. Hackl. A regularization framework for damage-plasticity models via gradient enhancement of the free energy. *International Journal for Numerical Methods in Biomedical Engineering*, 27:1199–1210, 2011.

[35] M. Dittmann. *Isogeometric analysis and hierarchical refinement for multi-field contact problems*. PhD thesis, University of Siegen, 2017.

[36] M. Dittmann, F. Aldakheel, J. Schulte, F. Schmidt, M. Krüger, P. Wriggers, and C. Hesch. Phase-field modeling of porous-ductile fracture in non-linear thermo-elasto-plastic solids. *Computer Methods in Applied Mechanics and Engineering*, 361:112730, 2020.

[37] M. Dittmann, F. Aldakheel, J. Schulte, P. Wriggers, and C. Hesch. Variational Phase-Field Formulation of Non-Linear Ductile Fracture. *Computer Methods in Applied Mechanics and Engineering*, 342:71–94, 2018.

[38] M. Dittmann, M. Krüger, F. Schmidt, S. Schuß, and C. Hesch. Variational modeling of thermomechanical fracture and anisotropic frictional mortar contact problems with adhesion. *Computational Mechanics*, 63(3):571–591, 2018.

[39] M. Dittmann, M. Krüger, F. Schmidt, S. Schuß, and C. Hesch. Variational modeling of thermomechanical fracture and anisotropic frictional mortar contact problems with adhesion. *Computational Mechanics*, 63(3):571–591, 2019.

[40] M. Dittmann, S. Schuß, B. Wohlmuth, and C. Hesch. Weak C^n coupling for multipatch isogeometric analysis in solid mechanics. *International Journal for Numerical Methods in Engineering*, 118:678–699, 2019.

[41] A. Dorfmann and R.W. Ogden. Nonlinear electroelasticity. *Acta Mechanica*, 174:167–183, 2005.

[42] R. Echter and M. Bischoff. Numerical efficiency, locking and unlocking of NURBS finite elements. *Computer Methods in Applied Mechanics and Engineering*, 199(5–8):374–382, 2010.

[43] A.C. Eringen. *Nonlocal Continuum Field Theories.* Springer, 2002.

[44] A.C. Eringen and E.S. Suhubi. Nonlinear theory of simple microelastic solids. *I. International Journal of Engineering Science*, 2:189–203, 1964.

[45] P. Fischer, M. Klassen, J. Mergheim, P. Steinmann, and R. Müller. Isogeometric analysis of 2D gradient elasticity. *Computational Mechanics*, 47(3):325–334, 2011.

[46] J. Fish. *Practical Multiscaling.* John Wiley & Sons, 2013.

[47] N.A. Fleck, G.M. Muller, M.F. Ashby, and J.W. Hutchinson. Strain gradient plasticity: theory and experiment. *Acta Materialia*, 42:475–487, 1994.

[48] S. Forest. Micromorphic approach for gradient elasticity, viscoplasticity, and damage. *Journal of Engineering Mechanics*, 135:117–131, 2009.

[49] S. Forest and R. Sievert. Elastoviscoplastic constitutive frameworks for generalized continua. *Acta Mechanica*, 160:71–111, 2003.

[50] S. Forest and R. Sievert. Nonlinear microstrain theories. *International Journal of Solids and Structures*, 43:7224–7245, 2006.

[51] F. Freddi and G. Royer-Carfagni. Regularized variational theories of fracture: A unified approach. *Journal of the Mechanics and Physics of Solids*, 58:1154–1174, 2010.

[52] M.G.D. Geers. Finite strain logarithmic hyperelasto-plasticity with softening: a strongly non-local implicit gradient framework. *Computer Methods in Applied Mechanics and Engineering*, 193:3377–3401, 2004.

[53] P. Germain. The Method of Virtual Power in Continuum Mechanics. Part 2: Microstructure. *SIAM Journal on Applied Mathematics*, 25:556–575, 1973.

[54] L. Greco and M. Cuomo. An isogeometric implicit G^1 mixed finite element for Kirchhoff space rods. *Computer Methods in Applied Mechanics and Engineering*, 298:325–349, 2016.

[55] A.A. Griffith. The Phenomena of Rupture and Flow in Solids. *Philosophical Transactions of the Royal Society of London A: Mathematical, Physical and Engineering Sciences*, 221(582-593):163–198, 1921.

[56] A. L. Gurson. *Plastic flow and fracture behavior of ductile materials incorporating void nucleation, growth and coalescence.* PhD thesis, Division of Engineering, Brown University, 1975.

[57] A.L. Gurson. Continuum theory of ductile rupture by void nucleation and growth, part I - Yield criteria and flow rules for porous ductile media. *Journal of Engineering Materials and Technology*, 99:2–15, 1977.

[58] C. Hesch and P. Betsch. Transient 3D Domain Decomposition Problems: Frame-indifferent mortar constraints and conserving integration. *International Journal for Numerical Methods in Engineering*, 82:329–358, 2010.

[59] C.B. Hirschberger. *A Treatise on Micromorphic Continua. Theory, Homogenization, Computation.* PhD thesis, Lehrstuhl für Technische Mechanik, Technische Universität Kaiserslautern, 2008.

[60] C.B. Hirschberger, E. Kuhl, and P. Steinmann. On deformational and configurational mechanics of micromorphic hyperelasticity - Theory and computation. *Computer Methods in Applied Mechanics and Engineering*, 196:4027–4044, 2007.

[61] G.A. Holzapfel. *Nonlinear Solid Mechanics.* Wiley, 2000.

[62] A. Huespe, A. Needleman, J. Oliver, and Sánchez. A finite strain, finite band method for modeling ductile fracture. *International Journal of Plasticity*, 28:53–69, 2012.

[63] G. Huetter, T. Linse, U. Mühlich, and M. Kuna. Simulation of ductile crack initiation and propagation by means of a non-local Gurson-model. *International Journal of Solids and Structures*, 50:662–671, 2013.

[64] T.J.R. Hughes, J. A. Cottrell, and Y. Bazilevs. Isogeometric analysis: CAD, finite elements, NURBS, exact geometry and mesh refinement. *Computer Methods in Applied Mechanics and Engineering*, 194(39–41):4135–4195, 2005.

[65] G. Hütter. Homogenization of a Cauchy continuum towards a micromorphic continuum. *Journal of the Mechanics and Physics of Solids*, 99:394–408, 2017.

[66] M. Ishaquddina, P. Raveendranath, and J. Reddy. Flexure and torsion locking phenomena in out-of-plane deformation of Timoshenko curved beam element. *Finite Elements in Analysis and Design*, 51:22–30, 2012.

[67] A. Javili, F. dell'Isola, and P. Steinmann. Geometrically nonlinear higher-gradient elasticity with energetic boundaries. *Journal of the Mechanics and Physics of Solids*, 61(12):2381–2401, 2013.

[68] M. Jeridi, L. Laiarinandrasana, and K. Saï. Comparative study of continuum damage mechanics and mechanics of porous media based on multi-mechanism model on polyamide 6 semi-crystalline polymer. *International Journal of Solids and Structures*, 53:12–27, 2015.

[69] J. Kiendl, K.U. Bletzinger, J. Linhard, and R. Wüchner. Isogeometric shell analysis with Kirchhoff–Love elements. *Computer Methods in Applied Mechanics and Engineering*, 198:3902–3914, 2009.

[70] J. Kiendl, M.C. Hsu, M.C.H. Whu, and A. Reali. Isogeometric Kirchhoff-Love shell formulations for general hyperelastic materials. *Computer Methods in Applied Mechanics and Engineering*, 291:280–303, 2015.

[71] J. Kinney. *Reinventing the Propeller: Aeronautical Specialty and the Triumph of the Modern Airplane.* Cambridge University Press, 2017.

[72] N. Kirchner and P. Steinmann. A unifying treatise on variational principles for gradient and micromorphic continua. *Philosophical Magazine*, 85:3875–3895, 2005.

[73] L. Laiarinandrasana, T. F. Morgeneyer, H. Proudhon, and C. Regrain. Damage of semicrystalline polyamide 6 assessed by 3d x-ray tomography: From microstructural evolution to constitutive modeling. *Journal of Polymer Science Part B: Polymer Physics*, 48(13):1516–1525, 2010.

[74] J.B. Leblond, G. Perrin, and J. Devaus. An improved Gurson-type model for hardenable ductile metals. *European Journal of Mechanics - A/Solids*, 14:499–527, 1995.

[75] T. Lehmann and U. Blix. On the coupled thermo-mechanical process in the necking problem. *International Journal of Plasticity*, 1:175–188, 1985.

[76] H. Li, M.W. Fu, J. Lu, and H. Yang. Ductile fracture: Experiments and computations. *International Journal of Plasticity*, 27:147–180, 2011.

[77] S. Love. The geoarchaeology of mudbricks in architecture: A methodological study from Çatalhöyük, turkey. *Geoarchaeology*, 27(2):140–156, 2012.

[78] A. Madeo, P. Neff, I.D. Ghiba, L. Placidi, and G. Rosi. Wave propagation in relaxed micromorphic continua: modeling metamaterials with frequency band-gaps. *Continuum Mechanics and Thermodynamics*, pages 1–20, 2013.

[79] J.E. Marsden and T.S. Ratiu. *Introduction to Mechanics and Symmetry.* Springer, 2003.

[80] G.A. Maugin. *Continuum Mechanics Through the Twentieth Century.* Springer Netherlands, 2013.

[81] G.A. Maugin and A.V. Metrikine. *Mechanics of Generalized Continua.* Springer-Verlag, 2010.

[82] A. Menzel and P. Steinmann. On the continuum formulation of higher gradient plasticity for single and polycrystals. *Journal of the Mechanics and Physics of Solids*, 48:1777–1796, 2000.

[83] C. Miehe, F. Aldakheel, and A. Raina. Phase field modeling of ductile fracture at finite strains. a variational gradient-extended plasticity-damage theory. *International Journal of Plasticity*, 84:1–32, 2016.

[84] C. Miehe, F. Aldakheel, and S. Teichtmeister. Phase-field modeling of ductile fracture at finite strains: A robust variational-based numerical implementation of a gradient-extended theory by micromorphic regularization. *International Journal for Numerical Methods in Engineering*, 111(9):816–863, 2017.

[85] C. Miehe, D. Kienle, F. Aldakheel, and S. Teichtmeister. Phase field modeling of fracture in porous plasticity: A variational gradient-extended Eulerian framework for the macroscopic analysis of ductile failure. *Computer Methods in Applied Mechanics and Engineering*, 312:3–50, 2016.

[86] C. Miehe, L.-M. Schänzel, and H. Ulmer. Phase field modeling of fracture in multiphysics problems. Part I. Balance of crack surface and failure criteria for brittle crack propagation in thermo-elastic solids. *Computer Methods in Applied Mechanics and Engineering*, 294:449–485, 2015.

[87] C. Miehe, F. Welchinger, and F. Aldakheel. Variational gradient plasticity at finite strains. Part II: Local-global updates and mixed finite elements for additive plasticity in the logarithmic strain space. *Computer Methods in Applied Mechanics and Engineering*, 268:704–734, 2014.

[88] C. Miehe, F. Welschinger, and M. Hofacker. Thermodynamically consistent phase-field models of fracture: Variational principles and multi-field fe implementations. *International Journal for Numerical Methods in Engineering*, 83:1273–1311, 2010.

[89] R. A. Mindlin. Micro-structure in linear elasticity. *Archive for Rational Mechanics and Analysis*, 16:51–78, 1964.

[90] R. A. Mindlin. On the equations of elastic materials with microstructure. *International Journal of Solids and Structures*, 1:73–78, 1965.

[91] R. A. Mindlin. Second gradient of strain and surface–tension in linear elasticity. *International Journal of Solids and Structures*, 1:417–438, 1965.

[92] R.D. Mindlin. Micro-structure in linear elasticity. *Archive for Rational Mechanics and Analysis*, pages 51–78, 1964.

[93] L.L. Mishnaevsky. *Computational Mesomechanics of Composites: Numerical Analysis of the Effect of Microstructures of Composites of Strength and Damage Resistance.* Wiley-Interscience, 2007.

[94] A.I. Murdoch and H. Cohen. Symmetry considerations for material surfaces. *Archive for Rational Mechanics and Analysis*, 76:61–98, 1981.

[95] K. Nahshon and J.W. Hutchinson. Modification of the Gurson Model for shear failure. *European Journal of Mechanics A/Solids*, 27:1–17, 2008.

[96] A. Needleman and V. Tvergaard. An analysis of ductile rupture in notched bars. *Journal of the Mechanics and Physics of Solids*, 32:461–490, 1984.

[97] P. Neff. The Cosserat couple modulus for continuous solids is zero viz the linearized Cauchy-stress tensor is symmetric. *Zeitschrift für Angewandte Mathematik und Mechanik*, 86:892–912, 2006.

[98] P. Neff and J. Jeong. A new paradigm: the linear isotropic Cosserat model with conformally invariant curvature energy. *Zeitschrift für Angewandte Mathematik und Mechanik*, 44:574–594, 2009.

[99] W.D. Nix and H. Gao. Indentation size effects in crystalline materials: a law for strain gradient plasticity. *Journal of the Mechanics and Physics of Solids*, 46:411–425, 1998.

[100] C. Polizzotto, G. Borino, and P. Fuschi. A thermodynamically consistent formulation of nonlocal and gradient plasticity. *Mechanics Research Communications*, 25:75–82, 1998.

[101] B.D. Reddy, F. Ebobisse, and A. McBride. Well-posedness of a model of strain gradient plasticity for plastically irrotational materials. *International Journal of Plasticity*, 24:55–73, 2008.

[102] J.M.L. Reis, T. de Santanna, P.P. Coutinho, A.H. Monteiro, S.C.S. Teixeira, E.G. Chaves, and H.S. da Costa Mattos. Coupled effect of ageing and temperature in the mechanical behaviour of a polyamide. *Polymer Testing*, 53:264–275, 2016.

[103] F. Reusch, B. Svendsen, and D. Klingbeil. A non-local extension of Gurson-based ductile damage modeling. *Computational Material Science*, 26:219–229, 2003.

[104] F. Reusch, B. Svendsen, and D. Klingbeil. Local and non-local Gurson-based ductile damage and failure modelling at large deformation. *European Journal of Mechanics A/Solids*, 22:779–792, 2003.

[105] K. Saanouni and M. Hamed. Micromorphic approach for finite gradient-elastoplasticity fully coupled with ductile damage: Formulation and computational aspects. *International Journal of Solids and Structures*, 50:2289–2309, 2013.

[106] J. Schulte, M. Dittmann, S.R. Eugster, S. Hesch, F. dell'Isola, and C. Hesch. Isogeometric shell elements for fiber reinforced composites. *Computer Methods in Applied Mechanics and Engineering*, 362:112845, 2020.

[107] J. Schulte, M. Krüger, M. Dittmann, and C. Hesch. Multi-field modeling of thermomechanical coupled fracture problems. *Proceedings in Applied Mathematics and Mechanics*, 2018.

[108] S. Schuß, M. Dittmann, S. Klinkel, B. Wohlmuth, and C. Hesch. Multi-patch isogeometric analysis for Kirchhoff-Love shell elements. *Computer Methods in Applied Mechanics and Engineering*, 349:91–116, 2019.

[109] N. Selles, A. King, H. Proudhon, N. Saintier, and L. Laiarinandrasana. Time dependent voiding mechanisms in polyamide 6 submitted to high stress triaxiality: experimental characterisation and finite element modelling. *Mechanics of Time-Dependent Materials*, 22:351–371, 2018.

[110] M. Silhavy. *The Mechanics and Thermodynamics of Continuous Media*. Springer, 1997.

[111] J.C. Simo and C. Miehe. Associative coupled thermoplasticity at finite strains: Formulation, numerical analysis and implementation. *Computer Methods in Applied Mechanics and Engineering*, 98(1):41–104, 1992.

[112] J.C. Simo and N. Tarnow. The Discrete Energy-Momentum Method. Conserving Algorithms for Nonlinear Elastodynamics. *Applied Mathematics and Physics (ZAMP)*, 43:757–792, 1992.

[113] K.P. Soldatos. Second-gradient plane deformations of ideal fibre-reinforced materials: implications of hyper-elasticity theory. *Journal of Engineering Mathematics*, 68:99–127, 2010.

[114] A.J.M. Spencer and K.P. Soldatos. Finite deformations of fibre-reinforced elastic solids with fibre bending stiffness. *International Journal of Non-Linear Mechanics*, 42:355–368, 2007.

[115] D.J. Steigmann. Theory of elastic solids reinforced with fibers resistant to extension, flexure and twist. *International Journal of Non-Linear Mechanics*, 47:734–742, 2012.

[116] D.J. Steigmann. Equilibrium of elastic lattice shells. *Journal of Engineering Mathematics*, 109:47–61, 2018.

[117] D.J. Steigmann and F. dell'Isola. Mechanical response of fabric sheets to three-dimensional bending, twisting, and stretching. *Acta Mechanica Sinica*, 31:373–382, 2015.

[118] D.J. Steigmann and A.C. Pipkin. Equilibrium of elastic nets. *Philosophical Transactions of the Royal Society of London*, 335:419–454, 1991.

[119] C. Steinke and M. Kaliske. Regularized variational theories of fracture: A unified approach. *Computational Mechanics*, 63:1019–1046, 2019.

[120] P. Steinmann. A micropolar theory of finite deformation and finite rotation muliplicative elastoplasticity. *International Journal of Solids and Structures*, 31:1063–1084, 1994.

[121] R.A. Toupin. Elastic materials with couple-stresses. *Archive for Rational Mechanics and Analysis*, 11:385–414, 1962.

[122] R.A. Toupin. Theories of elasticity with couple stress. *Archive for Rational Mechanics and Analysis*, 17:85–112, 1964.

[123] V. Tvergaard. On localization in ductile materials containing spherical voids. *International Journal of Fracture*, 18:237–252, 1982.

[124] V. Tvergaard and A. Needleman. Analysis of the cup-cone fracture in a round tensile bar. *Acta Metallurgica et Materialia*, 32:157–169, 1984.

[125] H.W. Wang, H.W. Zhou, J. Mishnaevsky, P. Brondsted, and L.N. Wang. Single fibre and multifibre unit cell analysis of strength and cracking of unidirectional composites. *Computational Materials Science*, 46:810–820, 2009.

[126] K. Weinberg and C. Hesch. A high-order finite deformation phase-field approach to fracture. *Continuum Mechanics and Thermodynamics*, pages 1–11, 2015.

[127] S. Wulfinghoff and T. Böhlke. Equivalent plastic strain gradient enhancement of single crystal plasticity: theory and numerics. *Proceedings of the Royal Society of London A: Mathematical, Physical and Engineering Sciences*, 2012.

[128] Z. Xue, M.G. Pontin, F.W. Zok, and J.W. Hutchinson. Calibration procedures for a computational model of ductile fracture. *Engineering Fracture Mechanics*, 77:492–509, 2010.

[129] F. Yang, A.C.M. Chong, D.C.C. Lam, and P. Tong. Couple stress based strain gradient theory for elasticity. *International Journal of Solids and Structures*, 39:2731–2743, 2002.

[130] U. Zdebel and T. Lehmann. Some theoretical considerations and experimental investigations on a constitutive law in thermoplasticity. *International Journal of Plasticity*, 3:369–389, 1987.

Schriftenreihe des Instituts für Mechanik
ISSN 2363-4936

Hrsg. von: Prof. Dr.-Ing. habil. Peter Betsch, Prof. Dr.-Ing. habil. Thomas Seelig

Band 1 Marlon Franke
Discretisation techniques for large deformation computational contact elastodynamics.
ISBN 978-3-7315-0278-4

Band 2 Philipp Hempel
Constitutive modeling of amorphous thermoplastic polymers with special emphasis on manufacturing processes.
ISBN 978-3-7315-0550-1

Band 3 Martin Helbig
Mehrskalenmodellierung von Schädigung in gummimodifizierten thermoplastischen Kunststoffen.
ISBN 978-3-7315-0571-6

Band 4 Maik Dittmann
Isogeometric analysis and hierarchical refinement for multi-field contact problems.
ISBN 978-3-7315-0616-4

Band 5 Yinping Yang
Numerical methods for the inverse dynamics simulation of underactuated mechanical systems.
ISBN 978-3-7315-0626-3

Band 6 Alexander Janz
Structure-preserving space-time discretization in a mixed framework for multi-field problems in large strain elasticity.
ISBN 978-3-7315-0926-4

Band 7 Mark Georg Schiebl
Thermodynamically consistent space-time discretization of non-isothermal mechanical systems in the framework of GENERIC.
ISBN 978-3-7315-1117-5

Band 8 Jonas Hund
Characterisation and modelling of PC/ABS blends.
ISBN 978-3-7315-1157-1

Band 9 Jonathan Schulte
Multi-field modeling and simulation of fiber-reinforced polymers.
ISBN 978-3-7315-1251-6

Die Bände sind unter www.ksp.kit.edu als PDF frei verfügbar oder als Druckausgabe bestellbar.